Essentials for Everyday
Email Writers

# 英文em@il
# 寫作溝通 的
# 第一本書

葉乃嘉
著

五南圖書出版公司 印行

# 推薦序

　　「溝」是種管道，用來疏通或調節水流，語言如同水一般，使用語言來傳遞訊息時，也需要有「疏通」，才不會因「溝渠」堵塞而造成不必要的誤會，因此，溝通技巧乃是我們所須面對的重要課題。書面溝通除了要求語意通順之外，還要求能精確傳達書寫者想表達的意思。可惜的是，如何有效溝通而不失禮節，常讓人摸不著頭緒。倘若文詞太過矯情，就怕無法寫出想要表達的重點；倘若太過直白，又怕因不得體而得罪對方，這當中該如何拿捏，乃是下筆者常常頭疼的地方。

　　東西方的思維不同，加上文化差異，西方人在陳述事情時，比較習慣開門見山，不太拐彎抹角；而華人在切入正題之前，則多會先鋪陳一番，假如不了解東西方的文化脈絡，就容易引起不必要的誤解。葉乃嘉老師的《英文E-mail寫作溝通的第一本書》，提供一個方便法門，讓人們在以文字交流時有脈絡可循。本書在溝通層面上同時考量了書信讀者和作者的立場，從分析受信者的心態開始，清理出正確的撰寫格式，再加上教導在寫信時如何開頭與結尾，讓書寫者容易表現得宜。

　　在網路發達的社會中，電子郵件雖然取代了紙筆書信文件，但不變的是文字的溝通技巧。良好的溝通，能讓寫信者把所要表達的資訊正確傳達出去，讓對方感受到其中所要傳達的意義，甚至感受到來信者的誠意。《英文E-mail寫作溝通的第一本書》為e世代人士提供了一個方便法門。本書從格式和溝通心理開始，讓讀者領略寫信時如何

開頭與結尾，在溝通時應具備甚麼態度，以及文字上適合使用何種技巧，章章都以精闢的論點和歸納的方式，提供寫作書信文件的方向，是您以文字與人溝通時可以參考的依據，也是一本不可多得的工具書。

陳瑞松

明道大學應用英語學系主任

2015年2月

# contents

Volume

# 1

# 書面溝通基礎

本篇提要

# 首　章
# 書信格式與溝通心理

　　第一印象常常占很大的分量，而外在美則是取得良好第一印象很重要的因素。

　　宜人的外觀是良好書信的第一要件。

　　引人注目的信讓人容易閱讀，而且看完會心情愉快。

　　讓人心情愉快的信比讓人感到乏味的信更能夠達到目的。

　　想要讓人對你的信有好的印象，信上就得先有容易讓人接受的特質。若是使用實體的信件，則信紙的品質要恰當，不宜用影印紙來充數，信箋要設計良好，以簡單樸素為原則。信的內容要簡潔親切，段落宜分明，內文的行距及間距應保持適當的比例，字裡行間以及邊幅的留白應該足夠，不必要的資訊只會使信箋顯得雜亂及擁擠，例如，分公司的住址、公司的產品、多餘的圖案……等，就屬於不必要的資訊，不要全放到信上。

　　促銷信件若能給人良好的第一印象，常常會搶得先機。在往訪問客戶前，書信可以做好鋪路的工作，讓客戶知道你往訪的時間和討論的主題，使雙方都有所準備，免於浪費彼此的時間。在與客戶會談後，也宜寫封信，把會談的重點作個總結，讓客戶有份具體的書面資料可以轉達給同事或上司。

 ## 常用的書信格式

　　常用的書信格式有好幾種，其中以*齊頭式*（Extreme Block

Style，見圖 1-1）最為簡單實用。此格式的特點是把每一段的開頭都自左邊邊線開始書寫，各行採單行行距，兩段間隔半至一行空行，插入表格時，表格與前後段均留半至一行空行。只要用熟了*完全切齊式*，其他的複雜格式可以不用多費工夫記憶。

　　英文公務書信的主要元素包括：信箋、日期、信內地址、敬稱、正文、信尾敬辭、簽名及職銜等 8 項（見圖 1-1），茲一一說明：

　　1. 正式信箋（Letterhead）：信箋是公務信件所應備，以簡樸的素色紙較為得體。信箋上端印有發信公司的下列訊息：

　　公司名稱及聯絡處。

　　電話及傳真或電子郵件信箱號碼。

　　其他選擇性的公司資訊。

　　若不使用公司信箋則可以將發信者之聯絡處書寫如下：

| | |
|---|---|
| 發信人名稱： | Mr. Naichia Yeh |
| 公司名稱： | YEH Publishing... |
| 門牌及街道名： | 1230 Fellowship Lane |
| 鄉鎮市名，省或州名： | Gaithersburg, MD 20878 |
| 國名（國內信可略）： | USA |

　　一般公務信箋都會註明公司的地址和電話，以利收信人在有必要時回信之用。但在大公司或政府部門，信箋上若只印了總公司或總機關的地址，你就有必要要在信上加註正確的回郵地址和電話，讓受信者知道該把回信寄到什麼地方。

　　即使公司的信箋上已列出所有分公司的聯絡地址與電話，你也該標明自己所在的辦公室，免得對方要回信的時候浪費時間捉迷藏。

**Information Technologies** ◀┈┈┈┈┈┈┈┈ 1
**No. 1234 Chuan-Yuan Road, Peitou, Taipei 00112, Taiwan**

July 20, 2005 ◀┈┈┈┈┈┈┈┈┈┈ 2

Mr. Barnaby Yeh,
Fictional Corp. ◀┈┈┈┈┈┈┈┈ 3
12330 Main Street
Gaithersburg, MD 20878, USA

Dear Mr. Yeh: ◀┈┈┈┈┈┈┈ 4

Thank you for sending your payment check on your April bill.

Our collection letters proceed automatically and occasionally a payment
crosses in the mail.  This is obviously what happened in your case. ┈ 5

Your check, of course, has been properly documented and your account
is currently marked paid in full.

Sincerely, ◀┈┈┈┈┈┈┈┈ 6

◀┈┈┈┈┈┈┈┈ 7

John Doe
Manager, Account Receivable ◀┈┈┈┈┈┈┈ 8

圖 1-1　英文公務書信格式

┈┈┈┈┈┈┈┈▶
識別記號　　　　　　　　　　　　Enclosure
◀┈┈┈┈┈┈┈┈
　　　　　　　　　　　　　　　　附件記號

**2. 日期（Date）**：格式繁多，但用下述格式即可：

June 20, 2005

**3. 收信地址（Delivery Address）**：受信者之姓名及地址，可書寫如下：

| | |
|---|---|
| 受信人名稱： | Mr. Sam Liu |
| 公司名稱： | International Techology |
| 路名及街道名： | 55 Chin-Hua Rd. |
| 鄉鎮市名，省或州名： | Peitou, Taipei 00112 |
| 國名（國內信可略）： | Taiwan |

**4. 敬稱（Salutation）**：除了 Dr.、Mr.、Mrs.、Ms. 外，不可採用簡寫，有關的稱呼頭銜一律用英文大寫，如：Mr. Ho, Dr. Chang, Sirs, Gentlemen 等。稱謂前冠以 *Dear* 不會有太大差錯。若收信者是陌生女性，以 *Ms.* 稱呼可以免去猜測其婚姻狀況的煩惱，但若對方有稱謂上的偏好，則以符合對方的期望為原則。

注意：不論中英文，連名帶姓地直稱對方都是極端無禮的，公務書信對受文者千萬要加上適當的稱謂。

**5. 正文（Body of Letter）**：段落宜簡短，內容應易於閱讀，將重點主題放在第一段，讓人可以一目瞭然，而且第一段應該要維持在二至三個句子，避免過於冗長，收信的人才會有往下閱讀的意願。

**6. 信末敬辭（Complimentary Closing）**：如 Yours sincerely, Yours faithfully, Yours truly 等。為了方便，可以一律用 *Sincerely yours*。

7. 信末簽名（Signature）：時至今日，除了簡短的傳真信函例外，已經難得有人用手寫英文書信，一方面手寫的筆跡不見得每個人都看得懂，另一方面，電腦文字處理程式及印表機已經成為每個公司所必備，即便是一人公司也不例外。既然信是印出來的，那麼親筆簽名也就格外重要，親筆簽名不但是一種認證，也是一種誠意的表現，如果不是電子郵件，建議在信件或卡片上，宜用親筆簽於信尾敬辭底下。如果數量實在太大，簽不勝簽，不妨請人代行，畢竟橡皮圖章不足以顯示你對該收信人的重視。寫信者不親筆簽名，難保會使某些收信者覺得不受尊重，因而延誤商機。

8. 職銜（Title）：職務的頭銜旨在表明發信者的身分，以世俗的眼光來看，當然職務越高的人所寫的信讓人越覺得重要。

至於電子郵件則比較不拘格式，一般可以省略前三項，因為電子郵件都會錄有寄件日期及寄件者的回信住址，e-mail 之信箋設計也甚為自由，一旦設計完畢，寫信時即自動出現，無須另外張羅，而簽名一項，可用直接打字或用簽名檔代替。

私信可以更不拘格式，可以省略一、八兩項，第四項也可以不用太正式，只要用 *Dear* 再加上收姓人的名字就夠了，但切勿連名帶姓。第六項則可以用 *Love* 或 *Yours* 等比較軟性的用字，若與對方不是很熟，那麼使用公務書信的格式也可以。

通常你在寫公務書信的時候，是代表公司寫信，因此適合用 *We* 來表示公司是個多於一人的組織，那封信若純屬個人的責任或代表個人的意見，就不妨用 *I*，在界線沒有那麼明顯的時候，用 *We* 比較不會有太多的爭議，這種原則用在代表家庭或其他組織的信件裡也同樣合用。

記住這個原則，代表自己說話時就使用 I，代表團體說話時就使用 We，舉例來說：

I am sure that we can arrange satisfactory terms.

　　意味作者個人（*I*）認為他能代表公司（*We*）協商出一個令對方滿意的條件。

　　又，由於西洋人的工作習慣使然，他們不愛在週末及其前後收到公務信件，因此在和西洋人業務往來時，應盡量讓信件在週二至週四期間到達，因為，他們在週五已經開始放假心態，而週一是剛放假回來，一般人在那兩天可能比較閒散，不太喜歡處理公務，類似心態在其他週休二日的地區也逐漸成形，因此也該同理視之。

## 書信寫作原則

　　想當然爾常是寫作者的盲點，作者自己不太容易察覺讀者的困難點，作者對自己所熟悉的題材，常有不切實際的同理心，認為自己寫的東西，別人要讀懂應該沒有問題，就像寫字潦草的人，不見得知道別人看不懂他的筆跡。

　　因此，虛心受教乃是增進品質的大好方法。

　　不要覺得不好意思，請別人來讀讀你寫的信，虛心接受他們的意見。請他們把看不懂的部分挑出來，作者習慣性忽略的地方，別人有時可以輕易看出，甚至連作者讀了好幾次也沒發覺的錯別字或文法失誤，別人也能一舉抓出。看信的人有時並不需要太專業，任何人只要有基本閱讀能力，都可以幫我們改善書信品質，因為別人不太會像作者一樣，把某些該說清的事想當然爾，輕輕帶過。

　　勤於檢查自己的作品，練習改善信中語氣，避免太過主觀或太過自我，把錯誤用色筆標示出來，刪去不必要的部分，改寫不當的文句，必要時重寫亦在所不惜，同時最好把修正稿保留，以供日後參考[1]。

---

1　最好利用文字處理程式裡的追蹤修訂功能，顯示出所有修訂的部分，存檔備查。

重溫自己的信之後，問問自己：

1. 你像是寫這信的人嗎？
2. 你願意收到這種信嗎？
3. 你願意回覆這種信嗎？
4. 你會喜歡寫這信的人嗎？
5. 你把這封信改寫得更好嗎？
6. 你和朋友是用這種方式交談的嗎？
7. 你能改進信裡的風格、語句和觀點嗎？
8. 這封公務信若公開宣讀，你會難為情嗎？

另外，讀讀別人的作品，看看它們：

1. 優點在哪裡？
2. 有哪些技巧值得借鏡？
3. 有哪些問題應該避免？

談到把信寫得更體貼、更友善，左欄這封催款信（collection letter）只能算是簡潔，談不上友善，把它改成像右欄那樣較為友善的寫法，成效更彰。

| 原信 | 修訂後 |
|---|---|
| Dear Mr. ＿＿＿: | Dear Mr. ＿＿＿: |
| You will agree that a collection letter should be brief and friendly to be successful.  Well, this letter is brief, and it is friendly. | Letter writing experts tell us that good letters should be brief and friendly in order to be successful.  That advice applies to collection letter, also. |

| 原信 | 修訂後 |
|---|---|
| Will you make it successful by sending your personal loan payment of $＿＿＿, which was due on June 1st without any further delay?  Please! | So I'm going to make this friendly letter brief and, I hope, successful.  Your payment of $＿＿＿ was due on ＿＿＿. Won't you please sent it, Mr. ＿＿＿? |
| Sincerely, | Sincerely, |

　　你與某人未曾謀面，一向只透過信件來往，你若想知道他對你的看法，只要讀讀自己以前寫給他的信就夠了，因為你大約已不太記得前信的內容，所以就能客觀地評量那位寫信者（也就是你自己）。這種方法會讓你像照鏡子一樣，看出別人對你的信有何觀點，信中一些你以前忽略的缺點，也會顯現出來。

　　本書提供了許多正面和負面寫作體例，為讀者建立起他山之石的情境，從其中看看別人的作品優缺點在哪裡，有哪些技巧值得學習，有哪些問題應該避免。

　　表 1-1 所列的書信寫作準則，是依英文字母次序排列的，這些原則對任何有意從事寫作的人都有幫助，也適用於包括中文在內的其他語言。

表 1-1　書信寫作原則 A － Z

| 原則 | 說明 |
|---|---|
| Attitude 態度 | 信會反映寫信者的態度，所以應確保語氣的愉快和友善。 |
| Brevity 簡單切題 | 迅速有禮地表明重點，不要旁敲側擊。 |
| Clarity 清楚 | 把信寫清楚，不要讓讀者費時費心來猜測你的意思。 |
| Discipline 有恆 | 持之以恆，多多練習寫信。 |
| Enthusiasm 熱情 | 以朋友的方式和對方在信上熱情溝通。 |

| 原則 | 說明 |
| --- | --- |
| Friendliness 友善 | 友善的書信較能帶來皆大歡喜的結果。 |
| Goodwill 善意 | 和氣生財，把每封信都當成是促銷信來寫。 |
| Humor 幽默 | 用幽默的態度讓你的信更生動。 |
| Imagination 想像力 | 想像自己是和對方隔桌對談。 |
| Judgment 判斷力 | 判斷什麼該說，什麼不該說，用什麼方式說。 |
| Knowledge 認知 | 知己知彼，認清自己，認清客戶。 |
| Language 語言 | 用顧客熟悉和喜歡的語言來交談。 |
| Modern 合時 | 揚棄陳腐的語句，用現代的語言來交談與書寫。 |
| Necessary 必要性 | 把必要的部分寫好，也別忘了讚美和祝賀等多數人都喜歡的一套。 |
| Organization 組織 | 精心安排和組織才來寫出好文章。 |
| Point of View 觀點 | 以讀者的觀點來敘述他們想要或需要知道的事。 |
| Quality 品質 | 儘可能強化信的品質。 |
| Readability 可讀性 | 內容宜流暢易懂，像對話一樣清楚明白。 |
| Sincerity 真誠 | 應該帶有發自內心的誠意。 |
| Tone 語調 | 友善的語調有時比內容更有影響力。 |
| Unity 統一性 | 信裡的內容該是你所想講的，不要旁生枝節。 |
| Vision 觀察力 | 將心比心，寫出讀者樂於一讀的東西。 |
| Willingness 意願 | 心甘情願，盡量服務讀者。 |
| X-ray 洞察力 | 知道何時要用什麼方式來說明什麼事。 |
| You-ness 以客為尊 | 多用 You，以讀者的觀點來取悅讀者。 |
| Zest 熱誠 | 用心讓讀者感受到誠意。 |

# 第一章
# 有創意的開場和結尾

本章提綱

◎應該避免的開場方式

◎值得參考的開場方式

◎用故事來做信的開頭

◎試試非傳統的問候語

◎可以參考的結尾方式

◎試試非傳統的結信語

你希望自己的信在一開頭就能令人印象深刻，那麼，到底要怎麼開頭？

「萬事起頭難」這種潛在想法總在偷偷控制我們的思考，進而影響我們寫信的方式，限制了我們所預期的效果。

改善的方法就是放鬆心情、簡單陳述事實，用友善的態度來表達，這樣一來，反而容易寫出好的文章。

 ## 應該避免的開場方式

書信有個好的開場會促進讀信人的心情，讓人樂於讀下去。下列有幾點是平常寫信應該注意的。

避免無意義的開場白。公務書信最好的開場方式就是在開頭說些會引起讀者興趣的事，開場白對信的文調常有很大的影響力，冷淡的開場辭會給人冷淡的印象，請看下例：

<u>This is to inform you that</u> it is a pleasure to send the enclosed check.

這句子的前六個字跟餘下的部分簡直就是前言不對後語，只要刪去了那六個字，就有了絕佳的開頭。

It is a pleasure to send the enclosed check.

再舉一個例子：

<u>We wish to say that</u> your order will be shipped Sept. 1

前五個字毫無必要，刪掉反而更好：

Your order will be shipped September 1

避免過時的語句。例如：

不　佳：Per your request re John Smith engaged in a phase of his engineering duties, we are forwarding same under separate cover.

修改後：**Here are two photographs showing Mr. _____ on his job, as you requested.**

不　佳：Re your favor of October 14, attached kindly find revised price list.

修改後：**Here is the revised price list requested in your letter
of _____ .**

避免不當斷句，要用完整的句子，例如：

不　佳：Confirming our telegram of today, formal proposal is
herewith enclosed.

修改後：**The enclosed formal proposal confirms our telegram
of this morning.**

不　佳：Answering your letter of the 5th, you will receive the
missing parts in five days.

修改後：**In five days you will receive the parts listed in your
letter of _____ .**

避免太突兀，要有連續性，例如：

不　佳：Re your letter of August 6th re employment with this
company.

修改後：**Thank you for your letter of August 6 expressing an
interest in a job with our company.（或 We are glad
to know from your letter of _____ that you are interested in working for our company.）**

不　佳：Per your request 6/19, are enclosing Bulletin #11

修改後：Bulletin #＿＿＿ is enclosed in reply to your request of ＿＿＿.（或 We are glad to send you a copy of Bulletin #＿＿＿ requested in your letter ＿＿＿.）

避免旁敲側擊，要迅速切入重點，例如：

不　佳：We are indeed most happy to have your fine letter of March 20th in which you state that you are interested in the details concerning the position which we are at the present time contemplating filling within the very near future.

修改後：Thank you for your letter of ＿＿＿ indicating your interest in the position as ＿＿＿.

不　佳：In reply to your very kind letter requesting some information regarding delivery on the above-mentioned order, we are pleased to be in a position to be able to inform you that said order will, in all probability, be ready and waiting shipment on or about September 14th.

修改後：You will be glad to know that your order No. ＿＿＿ will be shipped probably on ＿＿＿.

避免不必要的 *I* 和 *We*，例如：

不　佳：I am returning your manuscript enclosed, and I can assure you that I am profoundly impressed with it.

修改後：**Your manuscript, which is enclosed, profoundly interested me.**

不　佳：We wish to thank you for your recent inquiry.

修改後：**Thank you for your recent inquiry.**

避免使用過於僵硬的開頭，例如：

不　佳：This will acknowledge your letter of March 1$^{st}$ and inform you that we take great pleasure in sending you copy of our latest catalog.

修改後：**It is a pleasure to send you a copy of our latest catalog as you requested in your letter of March.**

不　佳：For your information, we can supply you with Gold, Red, or Rose immediately upon receipt of your order.

修改後：**We can ship you Red, Gold, or Rose immediately, whichever you prefer.**

避免用否定的方式開頭，若真有必要使用否定句，務必軟化語氣，例如：

不　佳：We cannot ship your order this week, but will send it next week for sure.

修改後：**Your order will be shipped next week for sure.**

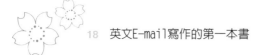
> 不　佳：You neglected to take your trade discount amounting to the sum of $_____ in payment of your order No. _____.
>
> 修改後：You will be glad to know that your prompt payment of order No. _____ earned you a discount of $_____, which has been credited to your account.

　　左欄這封信已經寫得很不錯了，但還是可改寫成像右欄那樣，讓它讀來像是寫給朋友的信：

| 原信 | 修訂後 |
|---|---|
| Dear Mr. _____: | Dear Mr. _____: |
| Enclosed you will find your copy of this week's issue of "Impact", which is one of the finer weekly newsletters covering current events and economic trends. | Here's a copy of the current issue of "Impact"—I think you'll enjoy it. |
| After making a rather extensive study of many similar publications, we found "Impact" to be a little more concise, a little more accurate, and that it had a little better presentation for the man on the go. | There are a number of publications like this, designed for busy people who want to keep abreast of things. We think "Impact" is the best of the bunch—more concise, more accurate, and a better presentation for the fellow who doesn't have time to burn. |
| Knowing that your time is valuable and that you are the type of person who likes to keep yourself informed on current national and business activities, we felt you would enjoy this publication. | This is only the beginning—you'll receive it regularly in the future, compliments of Bogus National. |
| As a result, we have arranged for you to receive this weekly newsletter with the compliments of the Bogus National, and we hope you will find it not only beneficial but also enjoyable. | Sincerely, |
| Sincerely, | |

## 值得參考的開場方式

好的開頭能引起讀者的注意或興趣。以下有十幾個方法，可以讓信有個成功的開始。

直接進入主題，例如：

1. The Christmas trees you ordered are being shipped today, Mr. _____.
2. The problem raised in your letter of _____ is not an easy one to answer.
3. Mr. _____ asked me to reply to your mail concerning late deliveries.
4. Your nice letter of June 15 was very much appreciated, Miss _____.
5. In your order, which was shipped today, two items were omitted.
6. Here are the specifications on which we are soliciting bids.

以問句作開頭，例如：

1. When may we expect shipment on our order of _____?
2. Have you reviewed your requirements for the coming season?
3. When do you expect to ship our order No. _____?

以認同收信者的方式起頭，例如：

You are quite right, Ms. _____. Conduct like that is not tolerated at The Regent.

用讚賞作為開場白，例如：

1. Your kindness is much appreciated, Mr. _____.
2. Thank you for your interest in our new models.
3. You are very thoughtful to forward the samples.

以客氣的請求開頭，例如：

May I trouble you for a minute, Mr. _____.

以禮物或建議作為開頭，例如：

1. The enclosed tickets to the _____ show are the compliments of the management, Ms. _____.
2. Please accept with our compliments the enclosed tickets to the Home Furnishing Show.

提及介紹人，例如：

1. Dr. _____, manager of _____, has suggested that I write you.
2. Your friend, Mr. _____, suggested that I write you.

若情況允許，可適時表達歉意，例如：

I want you to know how very sorry we are that you experienced that kind of service at our dealership, Mr. _____.

以適當的引言開頭，例如：

"Reach out and touch someone" has been AT&T's slogan for years.

使用新奇或令人訝異的開端，例如：

You don't owe us a cent, Mr. _____ —we certainly wish you did!

用友善的招呼語起頭，例如：

Congratulations on your promotion, Mr. _____! It couldn't have happened to a more deserving fellow!

使用節慶的賀語作為開端，例如：

Merry Christmas, Miss _____!

以事實的陳述起頭，例如：

We certainly appreciate your business, Mr. _____.

用一個特別的日子起頭，例如：

> July 27 is a very important date to both of us.

用手邊的具體事例作開頭，例如：

> Twenty of our home office salesmen are driving leased auto-mobiles at considerable saving.

用故事或者有特別涵義的事實來開頭，例如：

> My father had a favorite expression: "a hit bird flutters." Your recent letter caused such a flutter that I'm sure your criticism was justified.

適時醞釀親切的感情可以使整封信產生一種友好的氣氛，也是個好的開頭，例如：

> It's a pleasure to have an excuse for writing you. I still smile every time I recall the delightful evening we spent together at the convention.

用聳動的話題開頭，例如：

We can cut your utility costs by 10%—if not, our service costs you nothing.

 用故事來作信的開頭

好故事容易捉住讀者的注意力，所以用個好故事來作信的開頭是個好主意。

但要去哪找故事的題材呢？

很簡單！找些笑話，然後改編成你想要的故事。看看下面的例子：

In the early days of the Civil War, Abe Lincoln was deeply irritated by the indecisive, delaying tactics of General McClellan. "My dear McClellan," he wrote, if you do not wish to use the Army I should like to borrow it for a while."

In view of the repeated delays in filling your recent order, I wouldn't be the least surprised to receive a similar letter from you. We are, however, doing our very best in a difficult situation.

下面這個例子也不錯：

I saw a quote the other day from Ralph Waldo Emerson, something he wrote years ago: "A hero is no braver than the ordinary man, but is braver five minutes longer."

There's a lot of truth in it, Mr. _____, and it applies to sell-ing equally well. A great salesman isn't necessarily any different from an ordinary salesman except that he tries harder and keeps at is longer. That's what we need right now.

好故事不只是個給信開頭的好題裁，它也可以用在別的場合，比如說，故事可以幫你繞回自己想談的主題，在一封長信中，要變換主題或要重述先前的構想時，不妨引用一個新的故事。

其實，講故事這種方法，用在演講上也行得通，只要故事切題，不論用在信上或演講裡都大有好處。

 ## 試試非傳統的問候語

*Dear* 在英文信裡是一種制式的開頭用法，常被翻譯成親愛的，但在中文書信裡，有多少人會在寫信的時候稱呼受信人親愛的？

*Dear* 這個字只是傳統書信上的用詞，不具特別意義，在公務信件裡出現時，不免有些俗套，但它是在信的開頭中廣為使用的禮貌問候方式，不管為了什麼理由而捨棄了禮貌，都不可取。

除非跟一個人很熟，否則你不會上前去沒有任何問候就展開對話，至少該說聲「你好」，或介紹一下自己，不然，就是用其他禮貌的形式來引起對方的注意。總之，信頭沒有問候語是不禮貌的。

問候語對於整個信的品質很有幫助，你如果不用 *Dear*，就要想個替代的方式，千萬不要完全的忽略了問候。

用平常的招呼語或一般對話的方式來開頭，可以使信件更自然、友善和口語化，例如：

1. Thank you, Mr. _____! Your order for 25 cases of _____ arrived this morning.
2. No, Ms. _____ — we haven't forgotten your request.
3. Yes, Mr. _____, we can reserve a fine room for you July.
4. We're delighted, Ms._____, that you've chosen us to arrange your party.
5. Say, George, whatever became of that order you were going to place with us?
6. You're right, Ms. _____! We did make a mistake on your last order.
7. It's a pleasure, Miss _____, to add your name to our mailing list.
8. We've missed you, Mr. _____! You haven't placed an order with us since _____.
9. Last time I saw you, Mr. _____, you were talking about the possibility of _____.
10. Good morning, Dr. _____! Could I bother you for one minute?

　　上面這些例子只是提供一些新的開信語，當然用傳統的 *Dear* 來開頭並無不妥，只是寫信給較熟悉的客戶時，不見得要一成不變地以 *Dear* 來問候。

## 可以參考的結尾方式

　　結尾通常也可以給人很深刻的的印象，有力的結尾能使信件有畫龍點睛的效果，以下所列的方法，多少可以提供信件一個好的結尾：

　　以一個問句收尾，例如：

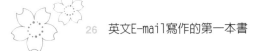

If you'd like more information, Mrs. _____, won't you let me know, please?

以一句**事實的陳述**收尾，例如：

You are always welcome whenever you stop by, Mr. _____.

以一個**保證**收尾，例如：

Remember, Jack, if you want to be sure, take my word!

以自己的**感言**收尾，例如：

Your compliments on our service are most sincerely appreciated, Mr. _____.

以**正面的陳述**收尾，例如：

Increase your word power and you will increase your earning power, Ms. _____!

在結尾處**表示感激**，例如：

It is a pleasure to add your name to our long list of pleased customers, Mr. _____.

用<u>銷售誘因</u>收尾，例如：

The hotel is only a 10 minute walk from 75% of the down-town shopping stores—a convenience you just can't afford to overlook, Miss _____.

用<u>親切和善意的祝福</u>收尾，例如：

Accept our best wishes for a great convention, Dr. _____.

用<u>適當的建議</u>收尾，例如：

Fill in and return the enclosed card promptly, Mr. _____, and your reservations will be assured.

用<u>節慶的賀語</u>結尾，例如：

Happy birthday, _____, from all of us!

使用任何<u>攀交情</u>的方式結尾，例如：

Next time you are out our way, Mrs. _____, be sure to stop in for a long chat and a few cups of that delicious coffee.

 試試非傳統的結信語

　　信件的結語可以不要太樣板化，下例中的替代句很可以用來取代慣用句：

慣用句：Looking forward to serving you.

替代句：**It will be a pleasure to serve you, Mr. _____.**

慣用句：Thank you again, Mr. _____.

替代句：**Your order is appreciated, Mr. _____.**

慣用句：If we can be of further service to you, kindly advise the undersigned.

替代句：**If we can help you further, please let me know.**

慣用句：Hoping the above has not inconvenienced you and awaiting your reply.

替代句：**I hope the delay has not inconvenienced you. We're awaiting your reply with great interest.**

慣用句：Thank you kindly for permitting us to tender you this quotation and trust we will be favored with your esteemed order.

替代句：**Thank you for letting us quote on this job, Mr. _____. If we get the order, you can be sure of prompt, efficient service.**

慣用句：I assure you it is my sincere and earnest desire to cooperate in every possible way in this and other instances.

替代句：**You can always be sure of our cooperation. Mr. _____.**

慣用句：Regretting our inability to serve you at this time.

替代句：**I hope next time we can be more helpful.**

　　至於公務書信結尾的敬辭，則可以一律選擇簡單的 *Sincerely yours*。

　　信件在簽名寄出之前，最好先讀一遍，如果沒有時間讀一遍就先擱下，暫且不要寄出。信末沒有署名是種嚴重的疏忽，有些人會認為這是嚴重的冒犯，如果收信人跟你很熟也就罷了，但若你寄出的是公務信件，收信人怎麼知道你是何方神聖。

　　寫作 e-mail 時也是一樣，你寫的既不是黑函，為何沒有署名？

　　沒再檢視一遍就寄出去的信，可能會對收信人造成不敬，也可能鬧笑話。請看下例：

Dear Mr. ＿＿＿＿＿ :

As you will probably recall, our little transaction last spring did not turn out satisfactorily for us. For that reason we do not care for any more of your business.

We will not ship your recent order. Thank you in advance for your future patronage.

Very truly yours,

　　「*Thank you in advance for your future patronage.*」和「*We do not care for any more of your business.*」出現在同一封信裡，豈不好笑？

# 第二章
# 萬無一失的溝通態度

本章提綱

◎親善的形象
◎愉快的心情
◎謙虛的態度
◎重視收信者的感覺
　把信的重心放在讀者
　從收信者的角度出發

　　e-mail 幾乎已經全面取代了傳統白紙黑字的書信，要寫出一封好的 e-mail，除了正確的拼字、完美的文法和良好的遣詞用句外，人際關係的運用也是不可或缺的一部分。

## 親善的形象

　　許多公務書信給人刻板的印象，這到底是怎麼回事？

　　也許是因為許多寫信者都誤以為：「公務書信既然是處理公務，就要用正式的、不帶個人感情的、客觀和條理的方式來寫作。」

　　但是，僵硬而不帶情感的公文或公務書信模式，絕不能算是良好的書信溝通方式。你雖在站在公務的立場寫信，但「公務」並非就該有張冷面孔。設身處地想一想，你是不是更樂於讀到一些人情味十足的信？如果是，為什麼不將己所之欲，施之於人？

　　只要你先表示誠意，大部分的人都願意友善相待，寫信的時候也

一樣，你若先釋出善意，對方也多會報以善意。不論是私信或公務信件，重點在於寫作時所用的心，在友善的心態下寫作的信，會引發讀者對你以及公司的愉悅感，進而造成愉快的互動。

每個人都有各自的人格特質，寫出來的信也該各有特色才對，如果你是有親和力的人，何妨配上親善的態度，用有親和力的方式來寫作。用愉悅友善的信來彰顯公司的風格，在客戶間建立鮮明清新的公司形象，絕對利多於弊。

好的書信除了應該具有善意，也要能夠引起讀者們的共鳴，即使是在傳達一個不愉快的訊息時，也可以藉著用友善、謙和的態度來清楚表達寫信者的意思，來避免傷害讀者的感情。因此，若把公務往來的書信都當成與朋友間的通信，用簡潔、真誠的方式，讓公務書信也具有私人信件般的感情，必能增加任務達成的機會。

以讀者的立場來看事情，比較容易寫出合情理的書信。

好信的先決條件在於信的本身應該具有：

1. 正確的語調和論點。

2. 正確的開頭和結尾。

3. 正確的主題和用詞。

4. 正確的應對態度。

能否靠信件促進業務的成長，很大的一部分在於信中給客戶的感覺是否友善。所以，在完成每封信件後，宜再加檢視，務必要親切與明白。

要寫出溫暖友善的信，自己就必須先有友善的心情。

揣摩一下作者寫下面這封信時的心情：

Dear Mr. _____ :

History tells us that many important things have happened throughout the years on ____/____. But my memory tells me that this day is important because it is your birthday.

That is why, before going home tonight, I am going to drink a glass of beer as a toast to you and when I do, I am going to wish you good fortune, good health, and the best of all the good things in life.

*A VERY HAPPY BITRHDAY TO YOU!*

Cordially,

　　這封信的作者態度自然又友善，能夠記得客戶的生日，更顯得細心體貼，接信的人肯定感到愉快。

　　建立親切的形象，對任何商家都是好事，請看下面這封信：

Dear Mrs. _____ :

Thank you for stopping in the other day. We are mighty glad that you found just the home furnishings you wanted.

If you ever look for something we don't have on display, just ask for it. If we don't have it in stock, chances are very good that we can get it for you...and we'd be more than happy to do it.

It was nice to serve you. Come back again soon, if only to browse around. We do appreciate your business and want to be of every possible service to you, Mrs. _____ .

　　這雖然只是封短信，收到的顧客卻可能有受到莫大尊重的感覺，因而對銷售者的周到留下深刻的印象，那麼，初次光臨時即使並未購買，也可能因而樂於來第二次、第三次，光臨的次數一多，購買的機會就相對提高。

　　左欄這封機械式的信顯得平凡呆板，雖然簡短但不夠友善，十足官樣文章。只要稍加用心，就能把信變得像右欄那樣，溫和親切，讓人感到溫暖。這兩封信表達的主題相同，真正的差別在它們所釋出的感情。

| 原信 | 修訂後 |
|---|---|
| Dear Madam: | Dear Ms. _____ 1: |
| This is to advise you that we have received two deposits to date which we have credited to your savings account #_____, as per your request. | It is a pleasure to tell you that two deposits totaling $_____ have been credited to your savings account. |
| Very truly yours, | Sincerely yours, |

　　信中的用語反映了你的想法、感覺和態度，感覺和態度總是互相影響的，你的信給客戶的感覺與該信的表達方式有關，而客戶對你的感覺則與那封信給他們的感覺有關，信上若有親切細心和體貼的感情，對方必然能夠感受。

　　友善又愉快的信絕對有效果。

　　下面幾封信中，第一封是某航空公司服務部門在接到旅客來信查詢失物時的回信：

1　知道對方的姓名，最好在信中給個稱呼，Dear Ms. Wong 畢竟比 Dear Madam 更有禮、更人性化些。

Dear Miss _____:

We have your letter about your lighter, lost on flight No. _____ of _____ Airlines.

Our Lost and Found Department is not located in Taipei, which is why we are unable to tell you whether or not your lighter was turned in, but we are checking.

If it is on hand, it will be mailed to you.  If you do not hear from us again, you can conclude that our efforts were unsuccessful. At this late date, the chances are slim, but we'll try.

Thank you for flying with us.

Sincerely,

　　第二封信航空公司找到失物後寫給旅客的覆信：

We have good news for you, Miss _____. Your lighter was found!

It will be mailed to you today or tomorrow and you should have it shortly.  Just thought you'd like to know in advance of the package.

Sincerely,

　　第三封是旅客感謝航空公司熱誠服務的謝函：

Dear Mrs. _____:

Thank you so much for your attention to the matter of my lost lighter. I am delighted that it was located.

Equally gratifying was the courtesy and enthusiasm, which seemed to pervade your two letters on what must be a rather routine matter for you. Such service is indeed worth a compliment.

Again, thank you for your help and success.

Sincerely yours,

上三封信展現了愉快的互動，是絕佳的書信示範。

某貿易公司負責人分別向三家電視台寫信，探問各家是否願以「售後抽成」的方式，爲這貿易商的產品打廣告。所謂售後抽成，就是電視台先免費播出廣告，待商品賣出後，從銷售額中抽取特定的百分比作爲回報。

以下是各家電視台給這貿易商的回信。

第一家電視台的回信：

Dear Sir：

I don't know where you got your information as stated in the opening sentence of you letter, "I have been informed that your station promotes items on a commission basis." And we don't appreciate it.

If you want to do business with us, buy; if you don't, don't write us.

Sincerely,

第二家電視台的回信：

Dear Mr. _____ :

This will acknowledge your recent letter offering your product on a commission basis. I am quite sure that you did not run your advertisement in the newspaper on a per inquiry basis. By the same token we will not accept business on a percentage or per inquiry basis.

For your information, I am attaching our rate card.

Yours very truly,

第三家電視台的回信：

Dear Mr. _____ :

Your information must have been somewhat misleading because KOTV has never promoted items on a commission basis.

We would be delighted to handle the _____, and I personally think it is an item of great demand.

As you possibly know, KOTV, channel 6, CBS basic, is in its seventh year of complete dominance in _____. You will find a rate card enclosed. The _____ Company will be delighted to furnish you with outstanding availabilities.

Thanks very much for thinking of KOTV.

Sincerely,

三家都不接受售後抽成的業務，其中第一家的回信更是近乎無禮，如果你是這個貿易商，終於決定做付費廣告，會考慮哪一家？有機會的話，又會推薦同行到哪一家去做廣告？

如果是我，答案絕不會是第一家。

而第三家電視台回信的方式婉轉，這一次雖沒賺到我的錢，但下次我大有可能因為他們的親切回應而尋求他們的服務。

在處理公務信件時，有人偶爾會想把「不上道」的對方給狠狠教訓一頓。如果你有這種衝動，千萬先忍住，不管對方再怎麼無禮，你沒有必要還擊，不妨用感性或幽默的方式來回應，所謂和氣致祥、和氣生財。

## 愉快的心情

把以前的信拿來重讀一次，看看自己寫信時的態度和心情如何。

信如其人，人的心情很容易反映到自己的作品上面，用快樂的心情寫信，那封信的正面感染力就變高，收信人的心情也不由得會好很多。

因此，寫信前不妨培養好心情，不管怎樣，千萬別在情緒惡劣的時候下筆。

放輕鬆，用友善親切的心，在信裡營造出感情十足的氣氛。寫封親切的信，關鍵不在文字的技巧，而是在於寫作的態度，請看下例：

It was a pleasure to have met with you and discussed the _____ System.

Should you desire further information, please feel welcome to call on me.  My telephone number is _____.

Thank you for your continued consideration, and looking forward to being of service to you.

　　這封信顯得簡潔有力又正式，文字上沒有瑕疵，然而還可以寫得再親切一點：

Thank you for your courtesy in discussing your mailing problems with me. They are interesting, and I was certainly impressed by your efforts to handle them in a systematic and orderly manner.

The more I think of your problems, the more I am convinced that the _____ Systems could provide valuable assistance. I'm looking forward to discussing with you again how we might help.

Meanwhile, if you would like further information won't you please call me at _____? It would be a pleasure to hear from you.

　　結論是，任何更人性化、更有趣的小技巧，都能使信更受歡迎、更有效。
　　請看這封動人的信：

Dear Mr. _____ :

As you can see from our name, we're a little country bank that specializes in pleasing small town and rural customers like you folks.

\_\_\_\_\_ Bank is one of the new breeds of local banks opened this Year. We are members of the FDIC and have their complete insurance coverage. Any time when you have any funds that you would like to put to work in a certificate of deposit, we would be very happy if you would contact your country cousins at \_\_\_\_\_ Bank!

Call us collect (\_\_) \_\_\_\_-\_\_\_\_.

寫下面這封信的人，給人的印象就是笑臉迎人，謙恭有禮又令人尊重：

TO OUR CUSTOMERS:

During the next two or three months, we will be converting gradually to a new computer system for processing orders, invoices, and shipping instructions.

Conversion to a new system always means confusion. But a new machine and a new system together could mean trouble. If any of you have had a similar experience, you probably know what we are talking about. If you should receive an invoice for $100,000 instead of $100, think nothing of it – our monster has probably goofed.

Seriously, everything in our power will be done to minimize ir-regularities. Some errors are going to get through in spite of our debugging program before the system will run smoothly. Please bear with us if one of these should come your way. Advise us promptly and we will rush corrected papers to you.

We are confident that as soon as we master all the applications on the electronic installation, the results will be beneficial to you and to us. In the meantime, please be indulgent – we can use you sympathy.

　　別人想到我們的好處會讓我們愉快，所以你想感謝、慶賀或恭維別人的時候，不妨寫個貼心的 e-mail 表達真誠的關心，這樣很容易打動人心。

## 謙虛的態度

　　個人或公司都應有謙虛的態度，不論是在公家單位回覆民眾的信件，或是為權威性的組織答覆客戶問題，都別忘了謙虛是美德，而正因為是代表大公司寫信，才更要分外掌握謙虛的原則，免得招惹怨謗。

　　左欄這封信有四段，每一段都以張揚自己的成就為重點，自誇得令人難以恭維。事實上，要人家認同自己的成就，用右欄的方式來寫才不那麼失當。修訂後的信以第一段標榜出自己的成就之後，就將重點轉了一個方向，說「本公司的努力受到政府的認同，希望貴客戶對本公司的產品品質和服務也很滿意」。少了許多自詡成就的成分，令人覺得不那麼可厭。

| 原信 | 修訂後 |
|---|---|
| Dear Mr. ＿＿＿: | Dear Mr. ＿＿＿: |
| It is our privilege to inform you that ＿＿＿Company, has just been awarded the President's "Efficiency in Export" award by the government. | Do you mind if we brag a little? As a customer you might be interested to know that the ＿＿＿ has just been awarded the President's "Efficiency in Export". |
| We are equally proud of the special citation that accompanied the award in which we were commended for our "significant contribution to the export expansion program". | It's nice to know that the government is pleased with our efforts abroad. We hope you and our other customers are equally satisfied with our products and service. |
| For more than 40 years it has been our policy to seek the most exacting standards in both our manufacture and in our sales and service activities. It has been our consistent policy to re gard the business given to us by our customers as a trust that must be met with these high manufacturing and service standards. | If there is anything we can do to improve them, I hope you will be the first to tell us. |
| That our Government has seen fit to place its stamp of approval on our work in the export field is most gratifying. Nevertheless, we regard the banner that we now so proudly fly, as more than an honor for work that has been done; we prefer to think of it as a challenge to us, a challenge to prompt us to raise even higher the standards that have permitted us to serve our customers so successfully over the years. | Cordially yours, |
| With our best wishes, | |

重視收信者的感覺

任何寫作的重點，都在於要能激起讀者的共鳴，或是把某種觀念傳達給讀者。

作品的成敗關鍵在於讀者，沒法把書信和報告寫得更好的原因，不全在寫作技術或能力欠佳，而可能在心態上沒有以讀者為本位。因此，不要太在乎自己的文采有沒有好好表露出來，而該多花精神考量讀者的接受度。

寫完一封信之後，在寄出之前檢視一下，想一想：

1. 這封信夠不夠友善？

2. 自己從這封信中得到了什麼？

3. 這封信是否讓我覺得我與收信者很投緣？

一、把信的重心放在讀者

在中文寫作中使用「我們」，隱約有「與讀者站在同一邊」的味道，但在英文書信中，有個與中文寫作不同的習慣，那就是：過度使用 *I*（我）和 *We*（我們）會少了「以客為尊」的味道，甚至會損害信中的語氣，因此應該盡量採用以 *You* 為主角的寫法。

左欄就是一封原本不錯的信被過多的 *I* 所損害例子。這封信中只有 83 個字，而 *I* 就占了 12 個。

1. 每 7 個字中便有 1 個 *I*，占全文近 15%。

2. 每個段落的開始便是 *I*。

3. 6 句中有 5 句以 *I* 開頭。

4. 只用了 2 個 *you*，占全文不到 5%。

右欄這封改寫後的信是以 *you* 做主角，在這封信裡：

5. 只用了 1 個 *I*，佔全文的 4%。

6. 段落的開頭並未使用 *I*，卻代以禮貌的 *Thank you*。

7. *You* 和 *I* 的比率為 5：1，是一個很好的示範。

　　在改寫後的信裡，讀者是主角，每件事是都導向他，而在字裡行間又稱呼了讀者一次，也是一個很好的用法，使該信顯得更體貼更友善。總之，應該多以讀者的立場來寫信，多用讀者的觀點來讀信。

| 原信 | 修訂後 |
|---|---|
| Dear Mr. _____ :<br><br>I was very glad to hear that the information that I sent proved useful. I am always glad to help a customer in any way I can. I am very sorry that I cannot send the other data requested, but this is confidential information, which I cannot release. I am sure that you will understand. I wish I would suggest some substitute, but I just can't think of any right now. However, if I do, I'll let you know.<br><br>Sincerely, | Dear Mr. _____ :<br><br>Thank you for letting me know that the information sent you was useful. It is always a pleasure to help a customer.<br><br>The other data you requested is, unfortunately, restricted and, therefore, cannot be released. Perhaps I can locate some unclassified material that will be helpful to you. If so, you will receive it promptly, Mr. _____.<br><br>Sincerely, |

　　左欄這封信長而無當，而且犯了過度使用 *We* 的忌諱，是個失敗的示範，有必要大刀闊斧地改寫。右欄就是改寫之後的例子，保留了該表達的重點，但顯得簡短友善得多。

| 原信 | 修訂後 |
|---|---|
| Dear Mr. _____ :<br><br>We wish to thank you for furnishing us the full names of the primary beneficiaries in connection with your recent application for life insurance with our company.<br><br>On _____ , you returned one of our memorandums indicating that the medical examination was taken May 13 and that we should receive the papers on the 17th.  As of this writing, the said examination has not as yet been received in our office.  We would like to suggest that you contact Dr. _____ or whoever completed the examination and thereby determine if it had been mailed to us when you said it was mailed.  If you should happen to find upon investigation that the examiner actually mailed the examination be completed immediately.  We most sincerely trust that this course of action will not be necessary for the examination, but that the original will be located in the doctor's office.<br><br>We wish to emphasize that we are indeed sorry that we are unable to give further consideration to your application, but as the medical examination is received we hasten to assure you that we will be more than happy to give prompt and immediate, attention to your application for insurance with our company.<br><br>Very truly yours, | Dear Mr. _____ :<br><br>Thank you for sending the full names of the primary beneficiaries of your proposed policy.<br><br>Your memo on _____ indicated that the medical examination had been taken on _____ and that we would receive the report in a few days. It has not yet arrived.<br><br>Would you please contact your examining doctor to see if he actually mailed us this report? If it has been lost, a duplicate will be necessary.<br><br>As soon as we receive this examination, your insurance application will receive prompt attention, Mr. _____ .<br><br>Sincerely, |

　　左欄這封信也過度使用了 We 和 Our，全文 54 字中就有 9 個 We 和 Our。右欄這篇則是那封信的改寫，全文變成以 You 為中心，有 8 個 you 和 your，而只有 2 個 we 和 our，從頭到尾都以讀者的立場來敘述，反映了良好的思考、寫作和銷售的技巧，好的書信就應該如此以客為尊。信中用公司的名字取代 Our Company，讓讀者多接觸了幾次你公司的名字，是很好的示範。

| 原信 | 修訂後 |
| --- | --- |
| Dear Mr. _____ :<br><br>We thank you for your recent inquiry about our products, and we are pleased to send you the information.  We trust it will be interesting and helpful.  Our building materials are sold by our dealers throughout the country.  We are taking the liberty of notifying our nearest dealer of your interest in our products.<br><br>Sincerely, | Dear Mr. _____ :<br><br>Thank you for your inquiry about WareEver products.  It is a pleasure to send you the information you requested.  We sincerely hope that this material will help you in your work , Mrs. _____ .<br><br>WareEver products are sold by 7-11 throughout the country.  Within a few days you will hear from Mr. _____ , manager of our Taipei office.  He will be glad to give you whatever additional information you may need.<br><br>Sincerely, |

　　要讓讀者更能接受自己的作品，就應該避免以下幾點：

1. 誇大其詞、言過其實。
2. 含意不清晰或太艱澀難懂。
3. 使用過時的商用詞彙或冷硬的樣板文字。
4. 忽略讀者的感受，用以上凌下的姿態*指示*或*教訓*讀者。

　　要檢驗自己有沒有這些毛病，就要重溫自己以前寫的信，依照上述幾個應該避免的原則，一一檢閱，然後記住千萬不要重蹈覆轍。

　　有些信件沒有在一開頭就顧到讀者，左欄這個就是這樣一封信的開頭段落，與其那樣用拉雜的事來開頭，不如把重心放在讀者的興趣上，而像右欄那樣直接講出重點：

| 原信 | 修訂後 |
| --- | --- |
| Dear Mr. ＿＿＿＿＿: | Dear Mr. ＿＿＿＿＿: |
| All business forecasts for the next year indicate steady holding of the present rate of activity with a possible general increase. To boost their profits from a marginal level many firms are reviewing costs of labor and material purchases. Some important items we suggest checking are your office and business forms. When is the last time that you reviewed their makeup, their layout and efficiency, and their cost?  Are you getting the best value for each dollar spent? | Outdated Business Forms Waste Money! |
| | When was the last time you reviewed the business forms you are currently using? Are they as up-to-date as the rest of your business?  How long since you've studied their layout and efficiency? Their cost? Do they represent the best value obtainable in today's market? |
| Sincerely, | Sincerely, |

　　有些寫信的人想用統計數字或公司的成就把讀者唬住，但是本書要強調的是，好的書信是要能一開始就引起讀者的興趣，而不是要引起讀者的佩服，若不能在開頭就抓住客戶的興趣，多半難以達到目的。

二、從收信者的角度出發

　　一封信成功與否，在於它是否足以表達寫信者的意思，而不在於信裡的文字多麼流暢，邏輯多麼有說服力。

　　許多人不善於體會讀信者的心理，就像左欄這封銀行通知客戶支

票跳票的信，忽略了一項很重要的人性，那就是：沒有人喜歡挨罵或受威脅，因此這封信不是很好的示範。要寫出好的信，需要善用心理學，不管這客戶是一時失察，還是有意詐欺，你還是該使用像右欄那樣親切而有建設性的方法。

　　若這封禮貌的通知仍然沒效，顧客還是有跳票的問題，那麼銀行還是可以發出禮貌的拒絕往來通知，不傷和氣。

| 原信 | 修訂後 |
|---|---|
| O.D. | O.D. |
| Those two letters printed in red on a bank statement indicate an overdrawn balance. You have no doubt noticed the frequency of this symbol appearing on your statement. We've noticed too—and with much concern. | Perhaps you've noticed these two letters in red on your bank statement. They've appeared so frequently lately we feel obliged to mention it to you. |
| We realize that once in a while an overdraft may be created through some error—a check written incorrectly in the register or perhaps not entered at all, or a mistake in addition or subtraction. An overdraft caused by error might occur in an account once or twice a year. The degree of regularity with which your account has been overdrawn leaves us little alternative but to consider the account unsatisfactory. Perhaps this situation may be corrected—we hope | O.D. indicates that your account has been overdrawn. It means you gave someone a check which exceeded the amount in your account. To save you embarrassment, we lent you the money and honored the check, rather than refusing to pay it and returning it marked "No funds". We're glad to extend this courtesy occasionally—anyone can make a mistake. Unfortunately, however, overdrafts are expensive and time-consuming to handle. |
| | Sometimes overdrafts are caused by lack of familiarity with the proper methods of balancing a checkbook. If this is the case, won't you stop in and let us explain them |

| 原信 | 修訂後 |
|---|---|
| so. If there is something you don't understand in the mechanics of maintaining proper records, please call on us. | to you? Perhaps you are trying to operate the account with too small a balance—an occasional personal loan might be helpful to prevent overdrafts at critical periods. |
| A week should allow sufficient time to place the account in order, don't you agree? A continuation of closing the account without further notice. | Whatever the reason, you'll find us anxious to cooperate to eliminate these O.D.'s and help operate your account on a mutually satisfactory basis. |

　　請記得要以讀者為主體，不要只顧著陳述事情而忘記顧及收信人的感情。左欄這封信就是個漫不經心的敗筆，僵硬、古板又不太友善，用了太多 *We*，好像是勉強拼湊的作品，因此可改寫得像右欄那樣，用正面的語調，以友善的方式起頭，以 *You* 縱貫全篇，少了許多本位主義味道的 *We*，充滿以客為尊的精神，而且篇幅少了一大半，簡短有力。

| 原信 | 修訂後 |
|---|---|
| Dear Sir: | Dear Mr. _____ : |
| In referring to your letter of April 23, we note that your letter was addressed to the _____ Bank. As you do not carry an account at this bank, we did not receive your letter as promptly as we should have. | Thank you for your letter of April 2. As soon as you sign and return the enclosed order for personalized checks, |
| As your letter of the above date was not signed by you, we do not feel that we have sufficient authorization to go | |

| 原信 | 修訂後 |
|---|---|
| ahead on our own and have printed up the personalized checks you apparently wanted.  Therefore we take pleasure in enclosing an order for personalized checks.  If you will kindly be so good as to fill out properly and SIGN the form attached, we will be more than happy to have the checks processed.<br><br>As of the present date, we are not holding any statements for you in our file.  We air mailed your March and April statements to you on May<br><br>Trusting we may be of service to you in the future.<br><br>Yours very truly, | we'll be happy to send them to you. Your signature is a formality, but a necessary one. Your last two statements were air mailed to you on May 10, Mr. _____.<br><br>Sincerely yours, |

　　以下再舉一個例子，某人因為某電視台新聞報導偏頗，寫信向新聞主管單位報怨，不久後，他收到一封制式的回信（見左欄），該信的敗筆在於強調自己每年要收到 50,000 封信，那人只是去信的 50,000 人中之一。

　　去信的人多雖是事實，但也不必直說出來，讓寫信者顯得微不足道。因此可改成像右欄那樣，體貼地點出相關條款，省去讀者自行翻找的麻煩。

| 原信 | 修訂後 |
|---|---|
| Dear Mr. _____:<br><br>Thank you for writing to the Commission. We receive approximately 50,000 comments, inquiries or complaints about broadcasting each year, and we believe that as a taxpayer you will | Dear Mr. _____:<br><br>Thank you for writing to the Commission.<br><br>I am enclosing a statement of FCC policies which apply to |

| | |
|---|---|
| understand why we try to reduce expenses by supplying a summary of some of the Commission's policies which we believe will answer many complaints and inquiries, including your own. | the questions you raised. You will be particularly interested in Sections 6(a), 7, 8, 10, and 1. |
| We appreciate your interest in writing to us, and if you believe that the material on the following pages does not adequately explain the Commission's policies in the area of your concern, we will try to provide a more specific answer. | We appreciate your interest in writing to us, and if you believe that the material does not adequately explain the Commission's policies in the area of your concern, we will try to provide a more specific answer. |
| Sincerely, | Sincerely, |

　　所以，我們不妨推己及人，偶爾用收信者的觀點來重讀自己的信，看看自己的感受如何。

# 第三章
# 文字溝通的藝術

本章提綱

◎小心無意的冒犯

◎從爲文中看個性

◎下筆要多帶感情

　道賀

　慰問

◎以正面代替負面

◎注重文字的表情

◎以肯定代替否定

◎善用 Thank you


　　寫信最重要的就是親切的口吻，不管是什麼信件，親切的口吻才能達到目的，硬邦邦的商業信早已不合時宜。

　　下面幾點也值得注意：

1. 不要質疑對方的動機。就算是銀行搶犯也不會覺得自己的動機有何可議之處，針對對方的動機發出評論，多半會激怒對方，即使對方眞的犯錯，你也要設想對方可能並非故意。

2. 批評別人前不妨自我檢討。如果你在某事上有該受批評之處，不妨先勇於認錯，自我檢討一番，隨後在你就同一件事批評別人的時候，對方也會比較少有怨言。

3. 要附和讀者：認清讀者群，採取他們較易接受的字、詞和用

語，迎合他們的風格。

## 小心無意的冒犯

有些遣詞用字容易引起不必要的誤會，所以不應該使用在書信上。下面所列的就是這類的文句，我們應該小心，以免無意間觸犯了別人。

1. If you actually did...
2. It is unreasonable of you to...
3. You must (should) have considered (known, understood)...
4. The error (mistake) you made...
5. We cannot (will not) accept (agree, believe, swallow) that...
6. We have been reliably informed that...
7. You are wrong in thinking that...
8. You do not appreciate (understand, value) that...
9. You failed (neglected) to...
10. You misinterpreted our intentions...
11. You overlooked...
12. You assert (believe, claim, say, state) that...
13. Your error (mistake)...
14. Your ignorance of the situation...

下面的字彙都屬於負面用字，也都在慎用之列：

| | | |
|---|---|---|
| at a loss | failure | offensive |
| annoyance | false | poor |
| bad | ignorant | serious mistake |
| careless | incorrect | terrible |
| complaint | inconvenient | unacceptable |
| disagreeable | inferior | uncalled for |
| disgruntled | misconception | unfair |
| disgusted | misled | unjust |
| dishonest | mistrust | unreasonable |
| displeased | misunderstand | unsatisfactory |
| dissatisfied | must | untrue |
| enlightened | negligence | worthless |

　　下例這封信就用了許多忌用的字，它不是憑空捏造，而是一封真正有人寫出來、寄到了對方手裡到的信，讀者不妨試試看，能不能把那些忌用字挑出來，然後好好改寫這封信。

We are at a loss to understand, your letter of _____ in which you claim that you are displeased and dissatisfied with the service which our company has tried to provide for you.  As you well know, Mr. _____, the premium, which provided coverage on your Buick was due and payable on _____, as of which date your policy, expired.  According to my records, you made no effort to reinstate this policy during the ensuing 30-day grace period.

We are, therefore, referring your complaint to our claim department, since we cannot accept your statement that you did reinstate policy #_____ within the grace period, as you claim. For her information, we are also sending a copy of both your letter and this reply to our Mrs. _____, your local service agent, for whatever action she may deem necessary.

We trust that this will satisfy you and assure you of our desire to be of service to you at any time.

　　左欄這封推銷信雖然寫得不錯，但是挑剔的人可能會覺得：此信暗示對方的印刷品質需要改進，可能會冒犯印刷部門的員工，所以不妨改成先誇獎對方的方式（如右欄）。該兩封信的差別並不在於文法上的優劣，而在後者比前者少了些冒犯的語氣。

| 原信 | 修訂後 |
|---|---|
| Dear Mr. _____ : | Dear Mr. _____ : |
| The printing on your recent circular indicates that your printing department needs Colorex Ink. We are pleased to send you a sample of our Black No. 123, along with our color book. | Congratulations on your recent circular! It's one of the finest pieces we've seen. |
| | Creative work and printing of this caliber deserve _____ Ink. We've enclosed a color book and a sample of our _____. Your offset department will be delighted with the excellent results using this formula. |
| We look forward to hav-ing you as our happy customer. | If you want more information or more sample, please call me. It would be a pleasure to serve you. |
| Sincerely, | Yours truly, |

　　要減少信中的瑕疵，就要在下筆時多斟酌，有些人提筆就寫，寫完就寄，難保不出些小狀況，部分收信人也許不在意細節，但較敏感的人卻可能因你信中某些用詞而不快。

　　請看左欄，注意到哪裡有瑕疵了嗎？

　　問題出在該信在要求而不是在建議收信人怎麼做，所以不妨把信改寫如右欄。

　　兩信的差別就在 *Please come...*（有要求性）和 *Would you like...*（只是建議性的），令人感受大大不同。

　　另外，左欄沒有開信稱呼（opening solutation），比較唐突而且缺乏感情，稍有貶低受信者的意味在裡頭。

| 原信 | 修訂後 |
|---|---|
| You recently paid off the account secured by __ ____ shares of _____. Please come into the bank some morning at your convenience to redeem this certificate. ...... | Dear Mr. & Mrs. _____: Now that you have paid off the loan secured by ____ __ shares of _____, we'd be happy to return the stock certificate. Would you like to pick it up at the bank some morning, or should we return it to you by registered mail? ...... |

　　人多喜歡別人以禮相待，因此用帶有感情的方式寫公務書信，可以花最少的力量獲得最大的成效，所以即使想簡化自己的信，也不要犧牲禮貌用語，因為禮多人不怪，這是人性。

## 從為文中看個性

從信中可以看出寫信者的特質,從一個人寫的信上觀察那人的個性,雖然不能完全準確,卻也八九不離十。

習於自誇的人,寫的信多半誇大不實。頭腦不清楚的人,寫起信來也是理路不清。寫信者若是幽默、機智又友善,你也可以從他的信上檢視出來。

所以說,你寫信的方式和內容會把你的個性和想法展露無遺。下面幾例子中就透露出作者的個性。

下面這封信閃爍其詞,寫信的人多半不怎麼可靠:

If you will act at once on this proposal, I am prepared to offer you a discount even lower than our previous low, low discount. Although you are a preferred customer and the kind of person we like to do business with, we cannot, of course, wait too long for your reply.

下面這封信文字纏夾不清,寫的人多半思路不清:

According to our records the material you ordered was dispatched from our _____ City plant on _____ via air freight and should have been received by you within ten days but in the event that a slip up in shipment occurred we will advise our _____ Department to initiate a second shipment and in the meantime if the original shipment does show up you can simply...

下面這封信值得取法，寫得友善隨和，很有誠意：

I'm sorry I was out when you stopped by the office on Wednesday, but, as _____ explained, I was at home nursing a heavy cold. Guess it's that time of year.

If you plan to be over this way again soon, why not be my guest for lunch and I can show you the new spring styles either before or afterward.

　　下面這封信是一家醫院的主管寫給病人的，寫信者充分了解病人的立場，給予病人一種值得信任的感覺：

Because illness is never welcome, it would not do for me to "welcome" you to _____ Hospital. But we do want you to feel that you well be among friends during your hospitalization— friends who are doing all they can do to speed your recovery. Everyone in this hospital, including those you may not contact, is interested in making your stay with us as pleasant as possible.

_____ Hospital is a non-profit institution operated to provide the best possible care for the people of this community. Only through the eyes of our patients can we see any shortcomings which might exist. If you have any suggestions or criticisms concerning any aspect of the care you receive, won't you please make them known to the nurse in charge of your department? Or, if you prefer, directly to me. In this way you will help us achieve our objective of providing continually improved service.

I hope your stay with us will be a short one and that you will help us to help you by giving us your comments and suggestions.

　　良好的書信表達方式不只一端，可能的話，你不妨試用上述的風格來改良自己或別人寫的信。

 ## 下筆要多帶感情

　　把笑容掛在臉上的業務人員絕對比吝於一笑者更得顧客歡心，愉快的促銷信件也比平淡的信要能多取得商機。仔細斟酌信中的主題，找出愉快和正面的寫法，表達得友善一點、笑容多一點，尤其在信的一開始，不要顯出負面的態度。另外，受信者若有任何值得讚賞的事，記得在信裡提出來，對方一高興，接受你建議的機會就高出許多。

　　下面列舉的就是一些帶感情的簡短句子，它們在人際交流的時候常常出現：

1. I love you.
2. I'm sorry.
3. You're beautiful.
4. Good work.
5. Thank you.
6. Let's get together.
7. You are so kind.
8. Can I help?
9. I need you.

　　類似的句子還有很多，這樣的寫法不僅省時省力，而且客戶也多半會認同。公務書信常見的缺點就是太冷淡了，不妨多用這類的短句來表達想法和感情。

## 一、道賀

　　適時使用賀卡、問候信、讚揚信和感謝函等，可能帶來商機。適時寄封發自內心表示讚許、欣賞、感謝的信，相信會給人帶來暖意，也可能贏取到不可預期的回報。寫這些信、卡或短訊雖要耗時花錢，但這一切都是值得的。

　　友善的短簡或短信可以讓受信人覺得窩心，花小錢用短簡來聯絡感情確實有效，也就是因為如此，在 e-mail 如此發達的今天，卡片生意仍未終結。

　　下面這封業務主管給屬下業務代表的簡短賀函，恭賀了對方的成就，也感激了對方在業務上的努力：

Dear ＿＿＿＿＿:

Congratulations for such a fine sale.　You should, indeed, be proud of such an achievement as the company now have an annual sales of $＿＿＿＿.

Thank you for your efforts in securing the ＿＿＿＿ account.　It has definitely helped to continue our success.

With best regards,

　　相信收信者一定會高興自己的努力受到肯定。

　　一封 e-mail 能間接替你做廣告，幫公司建立商譽、拓展商機、

鞏固業務，同時也贏取顧客的友誼，有可能讓你贏得一大筆生意，說不定成效會比花大錢做廣告來得好。在處理這麼重要的信件時，當然就值得花心思去斟酌。

　　良好的人際關係與公司業務息息相關，一個公司要有好的口碑，就須時時培養令人愉快的好印象，在小事情上多下功夫。

　　下面的這封賀函也許就能爲你的公司帶來一筆大買賣：

Dear Mr. _____ :

I read in yesterday's China Post that you have been elected President of the Retail Dealers Association. Congratulations, Mr. _____. Under your capable leadership, this organization will reach a new height in activity and service.

Sincerely yours,

　　賀函的要點是簡單、眞誠，只要坦白說出你的感受，對方就會有良好的印象。祝賀信寫來費時不多，而效果卻可能出奇地好，有時甚至短短幾個字就能贏得人心，甚至贏得商機。

　　但是短信的寫作切莫公式化或大量化，在大量製作的卡片上草草簽個名，大量寄出，多半是虛擲金錢。

　　每封信都是製造好印象的機會，這些好印象可能帶來意外的好結果，所以我們何妨把握每個結識朋友的機會，來創造有利的條件。

　　善於經營的人一有好念頭，就馬上寫下來分享給客戶，給客戶一個驚喜。例如，有位小姐在一家小店訂下婚紗，店主備妥婚紗之後，給小姐寄了封短信如下：

Dear Miss _____:

Just a note to tell you that your wedding dress is ready. We could send it, but I'm afraid if we did it would need pressing again.

I don't know when a gown and a girl have impressed me so. Lots of happiness, dear, and I hope we will see you soon.

　　這種親切的短信能給顧客帶來溫馨感，誰說公務書信就該有公務氣息？把形式主義給忘掉，開始寫些輕鬆、友善、人情味濃、真正會奏效的信。

## 二、慰問

　　慰問信要能簡單明白地表達真正的情感，而不需要華麗的詞藻，以下是兩個相當好的範例，其一：

Dear Mrs. _____:

I want to extend to you my sincere sympathy, and that of my associates here at _____ Company.

Although there isn't much we can say that will comfort you at a time like this, we do want you to know that Mr._____'s passing has made us feel we have lost a very dear friend.

If there is anything we can do to be of help to you, Mrs. _____, Please let me know.

　　其二：

Dear _____:

There is little anyone can say that will comfort you in your great loss, but I do want you to know that you have our heartfelt sympathy.

I had know and worked with your father for over thirty years. He was always good to me. His passing is a great loss to all of us.

這兩封慰問信沒有長篇大論，事實上，在這種傷感的情況又需要多說什麼？信愈長反而愈容易讓人感到不真摯，把想要表達的真情表現出來才是重點。

## 以正面代替負面

要先強調可以盡力服務的範圍，或能夠為對方效勞之處，而不要一開始就說自己能力所不及的地方。

以正面的語調取代負面的說明才是上策。

任何公司的業務代表在爭取客戶的時候都會客客氣氣、體貼有禮，但是，與銷售無關的事務部門就不見得會有同樣的好態度。左欄這封保險公司保單變更部門所寄出的信，就充滿了負面的態度，該信雖不是發自官方，卻是僚氣十足的官樣文章，一副「我們如果沒有收到正確的文件，就不會處理你的問題」的負面態度，令人不快。同樣一件事，可以用比較親切的方式，表示「只要我們收到正確的文件，就馬上處理你的問題」，如同右欄。

| 原信 | 修訂後 |
|---|---|
| Dear Mr. _____ : | Dear Mr. _____ : |
| Before we can consider the issuance of a duplicate policy, as requested in your recent correspondence, it will be necessary that the attached Affidavit of Loss be properly completed and returned to us for review. | I'm so sorry you lost your policy. Because it is a legal document, replacing it requires certain formalities I hope will not strain your patience. They are necessary, unfortunately, to keep the record straight. |
| You will note from the affidavit that we will require the signatures of two officers of your company as well as your own. These signatures must be witnessed by a Notary Public. Parts Ⅰ and Ⅱ of the form must also be completed. | Would you please fill out the enclosed Affidavit of Loss of Policy, including parts Ⅰ and Ⅱ ? You and two officers of your company should sign the affidavit and the signatures should be witnessed by a Notary Public. |
| When the affidavit has been returned to us, this matter will have our further attention.<br>Sincerely, | Thank you for bearing with us in what must seem like a lot of needless red tape. As soon as you return the affidavit, I'll see that a replacement policy is forwarded immediately.<br>Sincerely, |

　　請注意，不要先強調任何負面的訊息，或提到自己沒辦法為對方做到的事情。

　　某人寫信去兩家銀行，詢問關於開立新帳戶的事，第一家銀行回信表達了歡迎開戶之意，然後提到開戶的最低額度，同時可以用通信方式開戶：

Thank for your letter of _____, requesting a checking account for your son. A brochure on our bank and our affiliate is enclosed. We'll be happy to open an account at either bank. Please complete the form on the last page and return it to us with a check for the initial deposit.

Service charges are the same with both banks. If the balance drops below $_____ during the month, the charge is $_____ per check and a $_____ maintenance charge. Accounts with a $_____ minimum balance have no charge.

Please let me know if we can be of assistance prior to your son's arrival.

HAVE A NICE DAY!

第二家銀行的回信則以開戶最低額度作開頭，又講了一些服務費等會造成負面印象的事，還要求客戶親自到銀行去開戶：

In response to your letter concerning the opening of a checking account with our bank, on our personal account we require a minimum balance of $_____ for the entire statement period in order to bank free. If the balance should drop below this balance any time during the statement period there will be a $_____ service charge and an additional $_____ for any check processed during this time.

We order personalized checks from _____. These checks include your Name, Address and _____, if desired and may be purchased for approximately $_____ per order. The bank also prints checks that include just the Name and Account Number and may be obtained free of charge.

Upon your son's arrival, pleases have him stop by the bank so that we may fill out the forms needed for the proper handing of his account and order the style checks he desires.

We want to thank you for contacting our bank for your new checking account and are looking forward to meeting you personally. If at any time we can be of any further banking service, please do not hesitate to call on us.

　　比較這兩家銀行的覆信，想想看，你會選擇哪家銀行？
　　左欄是個負面取向的敗筆，如果能夠改寫成正面取向，有如有右欄，多半可以引起更多興趣：

| 原信 | 修訂後 |
| --- | --- |
| Springtime invites our attention to the out-of-doors. We were wondering if it had yours too. | In Spring, a homeowner's fancy turns to thoughts of |
| Springtime makes us more aware of the things around the home that need to be done: the new roof, the paint that has begun to fade, the patio that "would be | 1. a new roof<br>2. a fresh paint job<br>3. a family pool or outdoor patio<br>4. remodeling a few rooms or adding a new wing |

| 原信 | 修訂後 |
| --- | --- |
| nice" this summer and maybe even a swimming pool or a new landscape plan. All of these things are nice but they do cost money and sometime mean a cash outlay that is not in our savings accounts.<br><br>We realize that this is often the case, and we have therefore developed several programs, one of which we feel will provide the proper financing to meet your special requirements. | 5. attractive new landscaping<br><br>Whatever is on your mind, don't give it up for lack of money. Stop in and see us first. Many of these improvements are far easier to acquire than you might think—they can be financed by small, convenient payments over an extended period.<br><br>All it takes is a brief visit with one of our loan officers. He would be delighted to see you anytime. |

 注重文字的表情

　　什麼樣的方法能讓你的書信更有趣、更引人注意呢？

　　比較一下自己和別人寫的信，採取別人的優點，沿用一些會使人眼睛一亮的技巧和風格。

　　書信格調要是一成不變，會顯得寫信者不曉變通。何不拋下舊框框，用點新手法，讓別人的眼睛也一亮？加入肢體語言，會使演說更有趣、更明瞭、更有效。在書信裡也可以添加一些「肢體語言」，比如說，適當使用黑體、斜體、段落、底線和標點符號（逗號、句號、分號、破折號、引號、冒號、驚嘆號和問號等），會使書信更有趣、更易讀、更引人注意、也更具說服力。

　　請看看左欄這個「貴公司應該被恭喜」的、表錯情的例子，這封信的毛病就出在「*Your company is to be congratulated...*」上，這

個句子的語氣，讓人覺得寫信者自我膨脹得有點離譜，他好像居高臨下，有昭告大眾的權力，就像在說「本人決定頒獎給貴公司」一樣，實在不得體。不如把信改成右欄這個「恭喜您，某某先生……」的樣子：

| 原信 | 修訂後 |
|---|---|
| Dear Mr. _____ : | Congratulations Mr. _____ , |
| Your company is to be congratulated on its astute judgment in appointing you as the Manager of Equipment. Your interest, knowledge and zeal have always impressed me during the years it has been my pleasure to call on you. | On your appointment as Manager of Equipment. During the years I've called on you, I've had plenty of opportunity to observe your talents. It's good to know your company has recognized them. |
| Sincerely, | Sincerely, |

　　說到這裡，也許有人會發覺電視節目裡有時會誤用「頒」這個有上對下意味的字眼，主持人「頒獎」給參加者乃是「贈獎」之誤；同理，本國元首「贈勳」給外國公民，自然不應誤作「頒發」勳章。

　　另外一個常常誤用的詞就是「拜訪」，大家或者已經注意到，有些人偶爾會脫口說出像：

　　　我的朋友前幾天來拜訪我。

這類的話，殊不知「拜」有下對上的味道。

　　　我特地來拜訪你。

是一種客套用的謙語，沒有什麼不對，但是

> 請你有空來拜訪我。

就有了語病，即使是長輩或上司對晚輩或屬下，也都不適宜這樣說。

> 請你有空來坐坐（或聊聊、看看我等等）。

就比較得體。此外，「拜訪」一詞也有「探訪」一類的用詞可以轉用。

 ## 以肯定代替否定

即使是要提出否定的事，也要用正向的方式表達，才讓人較容易接受。左欄的四個例句較為冗長，且都是負面陳述，若是把它改為右欄的四個正面陳述，則句子較為精簡：

| 原句 | 修訂後 |
| --- | --- |
| 1. If you don't pay this installment by June 1, there will be an added penalty. | 1. If you make this payment by June 1, there will be no added penalty. |
| 2. This report never goes into any phase of the matter in detail, but covers each part briefly. | 2. This report discusses each step briefly. |
| 3. If the enclosed information is not sufficient and you need more, kindly let me know. | 3. If you need more information, please let me know. |
| 4. If it is not your desire to maintain both of your accounts, kindly advise us and we will be more than glad to transfer the funds in your checking account into your regular account. | 4. Would you like us to transfer the money in your checking account into your regular account? |

　　盡量以肯定的語氣來代替一些否定的用詞，這樣不但句子比較短，而且也比較容易達到目的。否定會讓人覺得沮喪，正面或肯定的字句則讓人感覺光明和快樂。因此，我們應學習積極地以肯定句代替否定句。像以下的例子：

負面：Due to the fact that printing charges are quite high, we have no other alternative than to make a slight charge of $_____ for this manual.

正面：**Although printing charges are quite high, you can still get this excellent manual for only $_____.**

負面：We regret that we cannot comply with your request regarding a supply of literature, as we are in the process of revising all our published data.

正面：**Our literature is being revised, Mr. _____. Just as soon as the printer delivers the new copies, we'll send you a supply.**

## 善用 Thank you

*Thank you* 兩字功效神奇，你可以從小事裡找到對方值得自己感謝的地方，即使實在找不出來，用 *Thank you* 也是禮多人不怪。

　　但是，使用這兩個字也是有場合及技巧的，通常可以使用下列的原則：

　　信頭──信的開始處可以用感激的話起頭，當然若有其它重要事宜亦不可忽略。

信體——*Thank you* 二字嵌於信中任何一個適當的位置都有很好的功效，此種用法較能顯出誠意。

信尾——*Thank you* 置於信尾也是常用的方法，但若信首已有此類字句，則可以不用重複。

上述三項只是原則性的，重要的是要讓讀者感受到你的感謝語是真誠的，如果不是出於真心，那麼這個 *Thank you* 不用也罷。下面這個例子就是一封在信末勉強加上 Thank you 的敗筆。

Regarding the deferment of the $_____ payment due on you note October 14, we cannot grant you this extension and shall expect your payment when due. Thank you!

但下面這封感謝函也許馬上就會讓顧客再度上門：

Please accept this thank you for being one of our best customers this year.

We appreciate your business and hope to continue serving you during the coming year.

第四章

# 用字遣詞的技巧

本章提綱

◎應該簡明直接
　　多用簡單詞彙
　　力求口語化
　　盡量使用主動語態
◎避免長而無當
　　理出表達的重點
　　避免不必要的重複
　　刪除累贅的字與詞
　　直截了當恰到好處
◎避免措辭公式化
　　揚棄陳舊文字
　　不要官腔官調
　　變換表達方式
◎訴求宜明顯統一

　　沒有語言和文字，人類無法把所知道的傳給後代，自然也就無法靠知識的累積來發展文明，因此，語言和文字可以說是人類最偉大的發明。人類利用語言文字來交換思想，所以理解和使用語言文字的能力，是成功的一個重要因素。

看看知名作家 *William Styron*[1] 的文字片段：

> Helen closed the door.　In her room everything was sunny and clean. A soft breeze shook the curtains; they trembled slightly, as if with the touch of a feeble and unseen hand. Outside the window the holly leaves rustled, made thin, dry scrapings against the screen, and then this breeze, so familiar to her because of its nearly predictable comings and goings, suddenly ceased: the curtains fell limp without a sound and the house, sapped of air, was filled with an abrupt, wicked heat, like that which escapes from an oven door.

　　好作家鮮少用冗長的詞句來表達思想，他們只用簡單的字彙就足以生動地敘事或抒情。如果他們選用複雜的字彙，那麼讀者連要搞懂那些字彙都有困難了，怎麼還有餘暇去欣賞他們的文章呢？

## 應該簡明直接

　　信要能夠明白易懂才能收效，但是要怎樣把信寫得清晰易懂呢？下面這封信提供了很好的答案：

---

[1]　當代美國作家（1925-），其所著的 *Sophie's Choice*《蘇菲的抉擇》與 *Darkness Visible: A Memoir of Madness* 分別列名於藍燈書屋（Random House）當代文庫（Modern Library）20 世紀百大英文小說（100 Best Fictions）及百大英文非小說（100 Best Nonfictions）之中。

It's easy, Mr. _____, to put personality into a shout letter.  Just relax.

A good way to relax is to use an opening like the one above rather than the formal Dear Mr. _____.

This kind of opening also helps get directly to the point—that's particularly valuable in a short letter.

Think of the person to whom you are writing as a real, live human being—he or she really is!  Picture your reader as sitting next to your desk and write in a natural manner, using the same words that you would say if the person were actually sitting there.

Don't worry about how short your letter is.  As long as you've answered the question and close with a pleasant thought, it won't be curt.

Hope it's as nice a day in Taipei as it is here!

Sincerely,


溝通的目的是要讓人了解，而不是令人費解。


一、多用簡單詞彙

　　完全摒棄長字當然沒有必要，只是要知道，長字會使你的信比較無力。

　　管理階層要做好管理工作，在溝通的時候就要使用廣泛「被管理者」所能了解的語言文字。不論訊息是給學者或是給識字不多的人，

簡單的詞彙能讓溝通更容易。

　　英文是由許多種文字所組合而成，當中有許多詞彙不是一般人所知。下面兩個句子，第一句一看就懂，而第二句雖然意思相近，但也許還會有人傻傻地問：「你是要我走嗎？」

---

1. Please get out of here!
2. Please remove yourself from these premises immediately!

---

　　你的客戶若身屬非英語系國家（像是亞洲各國），他們可能難以了解你精工雕琢的「典雅文字」，就算他們在費時間查字典之後終於弄懂你的意思，該信的效力也已經大打折扣。因此，請不用在信中顯示自己的文學素養。

　　充分的詞彙有助你了解他人，也能助你傳達自己的想法，但是以不常用的詞彙來寫信是不智的，太艱澀的詞彙容易造成溝通障礙。

　　簡短的詞彙比冗長的詞彙更富情感、更有影響力，可以使信不那麼冗長乏味，也讓事情比較清楚易解。

　　看看左欄這份某公司高級主管給員工的備忘錄，是不是有點難以了解？員工的教育水準不一，何妨把這備忘錄改得像右欄那樣簡單明白些：

| 原文 | 修訂後 |
| --- | --- |
| 　　The varying incidence of holidays is an unmitigated nuisance in scheduling industrial production.<br>　　Those which fall at the extremities of the week cause less interruption of output than those which occur in the middle. | 　　The fact that holidays fall on different days of the week makes it difficult to plan factory operations.<br>　　When holidays fall on Monday or Friday—making a three-day weekend—we lose less production than when a holiday comes in the middle of the week. |

這是個選舉的時代，如果候選人還在文宣品裡頭使用文言文，拉票的效果應該不會很大。

長字化簡表所列的是一些長字和它們清晰有力的代用字，大家可以盡量參照使用。

## 二、力求口語化

許多人閱讀的時候是用默唸的方式，他們傾向於把字一個一個地通過腦子默讀出來，因此，口語化的文字較能有效進入他們腦中，反之，非口語化的寫作就不那麼容易被吸收。

你可以將一些「正式」和「嚴肅」的文字口語化之後，把兩種都試讀一下，自行證明。

若你還認為公務書信必需正式而嚴謹，那要請問，通信時為什麼要和對面交談時使用不同的語言？寫信時有充分的時間思考和修改，不太會犯跟面對面說話時一樣的語病，輕鬆一點不妨。

左欄是一封銀行給顧客的信，不妨按照「放輕鬆」的原則，將信改寫如右欄：

| 原信 | 修訂後 |
| --- | --- |
| Dear Mr. _____ : <br><br> Attached herewith is an advice of charge of \$_____, which amount represents the closing of your account with us and the transfer of said amount to our Taipei branch to be used for opening a new | Dear Mr. _____ : <br><br> The attached statement shows that your balance of \$_____ has been transferred to a new account in your name at our _____ branch. It has been a pleasure to serve you at this branch, Mr. _____, but we realize that the Taipei branch will be more convenient for you. Mrs. _____, the manager, would like very much to meet you. |

| 原信 | 修訂後 |
|---|---|
| account in your name.<br><br>Sincerely, | Won't you please speak to him the first time you are in the bank?<br><br>Sincerely, |

　　下文左欄是封求職信的第一段，88 個字裡就有 35 個字（41%）是多於 1 個音節的。此外，文章本身不但不口語化，而且還纏纏夾夾，難以卒讀。

　　既然書信以口語化為宜，何妨像右欄一樣寫得簡單明白：

| 原信 | 修訂後 |
|---|---|
| Dear Mr. ＿＿＿＿＿:<br><br>I am writing to you directly concerning the possibility of employment with your company. At the present moment, I am pursuing a Ph.D. program at ＿＿＿＿＿ University, but now having passed the M.A. stage of the program, I am hesitant as to the efficacy of further committing myself to academic work as a fulfillment of my basic career objectives. The following cursory description of my personal history and statistics should better enable you to judge my qualifications for any possible openings which your bank might have.<br><br>Sincerely, | Dear Mr. ＿＿＿＿＿:<br><br>I am writing to you for a possible position with the ＿＿ ＿＿＿＿. While having passed the M.A stage and studying for my Ph.D. at ＿＿＿＿＿ University, I am thinking to change my career goal instead of continuing my academic work. The record below should better describe my employment qualifications.<br><br>Sincerely, |

　　生活在現代就要用現代的寫法，過時的寫法已經沒有存在價值了。

　　舊詞口語化表是一些過時的和現代寫法的對照，說穿了，現代寫法

就是對話式的寫法，對話式的寫法使信多了一份人性，少了幾分枯燥。

三、盡量使用主動語態

在中文裡原本很少使用被動語態，但由於受到西潮的影響，中文裡出現被動語態的情形也就增多了。被動語態常出現在老式的英文裡，曾經是一種流行的寫作方式，但經過有識之士幾十年來的宣導，已經使被動語態逐漸失去市場。

主動語態比被動語態簡短有力，也能夠更清楚、更直接地表達文意，比較下面兩個同義的句子，就可以分辨出孰優孰劣：

1. A decision was made by the committee...
2. The committee decided...

因此，在任何寫作中，若不是要刻意經營出一種被動的態勢，就該盡量避免老舊僵化的被動式。

 避免長而無當

2 個音節以下的字，通常比多音節的字容易閱讀，因此要讓人願意讀你的信，就不妨：

1. 盡量把句子縮短在 15 個字以內。

2. 多使用兩個音節以下的字。

日常對話常用字中，大部分只有 1、2 個音節，這些字容易了解，容易發音，讓人印象比較明確。因此，有同義字可選時，只要不會使文意偏離或誇張，就該盡量選用音節短的那一個。

長句分析起來很不容易，會讓人讀到了後面就忘掉了前面，很不

可取。

　　讀完左欄那長達一段的句子，看你能了解多少，再看看右欄修改後的例子，把原文斷成四句之後，是不是容易懂得多了？

| 原文 | 修訂後 |
|---|---|
| As you know, your year's guarantee is one which covers defective workmanship or materials which would reveal itself under normal use conditions within twelve months, likewise, it exempts the company from mechanical failures due to accident, alteration, misuse or abuse, and time also accounts for a certain amount of deterioration which obviously cannot be covered by our guarantee. | Your one-year guarantee covers defective workmanship. It also covers materials under normal use conditions within 12 months. The guarantee exempts the company from mechanical failures due to accident, alteration, misuse or abuse. Certain amount of deterioration through time obviously cannot be covered. |

　　長句容易造成文章冗長、閱讀耗時，短句則較易理解。但短句子當然不是萬靈丹，有時候重組一下短句子，能把意思變得更清楚。

　　下面這包含了三個句子的範例簡單易懂：

　　The main ingredient in natural abrasives is crystalline aluminum oxide. Emery was the first natural abrasive found on earth. It is 50% crystalline aluminum oxide.

　　而改成如下的兩個較長句子也不差：

> The main ingredient in natural abrasives is crystalline aluminum oxide. Emery, the first natural abrasive found on earth, is 50% crystalline aluminum oxide.

　　請記住，文字的目的在達意，不在考驗讀者的分析能力，你的目的不是選短的字，而是選對方容易了解的字。

## 一、理出表達的重點

　　有些執行階層的人士慣於使用華而不實、長而不當的寫信方式，原因之一是懶，因為那種方式已經成為一種標準形式，可以隨時套用，不太需要思考。另一個原因，是那些人認為只有這種方式才適合大公司，簡單明瞭的方式只適用在非知識階層，但是他們錯了。

　　有些人誤以為寫給重要客戶的書信，就該用重量級的詞彙才能表示重視對方。事實上，越重要的信越要用簡明、直接的方式來寫，以免讀者誤會了信的內容。

　　長信有時雖不可避免，但就公務書信而言，簡潔易讀的信不但寫來經濟，且在傳達重點時也不會有雜訊來引開讀者的注意。

　　像左欄這封信就長而不當，寫信者沒打草稿，寫了半天都點不出重點，是個失敗的例子。這封信至少應該表明下列三點：

　　1. 指出雙方的誤解。
　　2. 澄清誤會。
　　3. 避免造成對方的反感。

　　根據這些重點，原信就可改寫成右欄得那樣簡單明瞭。信短了一半，卻表達得更為清晰，所以你若想把信寫得有條理，就別怕麻煩，先打草稿或列出重點，再一步步把信架構出來。

| 原信 | 修訂後 |
|---|---|
| Dear Mr. _____ : | Dear Mr. _____ : |
| We have received a statement from you for what appears to be storage for a nine-month period up to December 1st, which appears to be at the rate of $_____ per month. | When you agre-ed with Mr. _____, our representative, to handle our feeds, he told us that your storage charge would be $_____. |
| We wrote our Mrs. _____ who has served you in the past, to contact you and determine the extent of our bill in view of the fact that he had informed us that the amount of storage charge would be at the rate of $_____ per ton of feed, which was handled by you. | Does this agree with your recollec-tion? Your storage bill for the nine-month period up to December 1 puzzles us because it appears to be at a flat rate of $_____. |
| In view of the fact that Mr. _____ states the rate would be at the rate of $_____ per ton of feed, we are wondering if this agrees with your information about the arrangement you made with Mr. _____ for handling our feeds shipped in that area to customers of Ms. _____ . | |
| We very much appreciate having the opportunity to work with you on this and on other matters, Mr. _____ . Also, we greatly appreciate the business you have given us in the past. | If you will tell us your recollection of the agreement, I'm sure we can get this matter straightened out. We certainly appreciate the job you are doing for us. |
| Our problem at the moment seems to be a misunderstanding as to the nature of our arrangement with you for handling our feed. | Sincerely, |
| We would appreciate hearing from you as to whether Mr. _____ stated anything to you regarding a storage rate per month since he has told us that his arrangement with you was on the basis of the feed you were to handle for us, at the rate $_____ per ton. | |

| 原信 | 修訂後 |
|---|---|
| Thanking you for your cooperation in this matter, and hoping to hear from you at your convenience.<br><br>Yours very truly, | |

請記得，你的信顯得長而無當的時候，就是該大力刪減的時候。

## 二、避免不必要的重複

有些人為了加強語氣，愛在同一句子裡把事情說兩次，用的雖然不是同一個字，但仍犯了重複的毛病（見左欄）。多餘的重複不見得有效，相形之下，精簡的陳述會比較有力，因此可把這些句子改寫如右欄：

| 原句 | 修訂後 |
|---|---|
| 1. First and foremost, let me say that your order has been completed and will be shipped today. | 1. First, your order has been completed. It will be shipped today. |
| 2. Each and every one of them has been carefully tested. | 2. Every one of them has been carefully tested. |
| 3. We wish to express our sincere and earnest regret. | 3. We wish to express our sincere regret. |
| 4. We think your decision is unjust and unfair, Mr. Smith. | 4. We think your decision is unfair, Mr. Smith. |
| 5. Isn't this a fair and reasonable adjustment? | 5. Isn't this a fair adjustment? |

有些人認為用長而艱澀的方式敘事，才能顯出自己的才學，所以他們寫信時就會用自己所知道的詞彙一再重複相同的事，寫成左欄的

例子那樣。其實像右欄那樣把信寫短些，反而可以把事情說得更好：

| 原信 | 修訂後 |
|---|---|
| Dear Mr. _____ : | Dear Mr. _____ : |
| In reply to your kind favor of March 23, relative to information concerning your graduate, Mr. _____, and the work that said graduate is currently engaged in with our company, we hereby inform you that said Mr. _____ is now Private _____. | Thanks for your nice letter of March 23 asking about Mr. _____ and what he is doing in our company. |
| Mr. _____ was drafted into the Army of the United States somewhere in the neighborhood of last September 15th. Around the first of March, this year, we heard from him form _____. He is making it a point of keeping in touch with us from time to time as a matter of good public relations policy. | Mr. _____ was drafted last September. He was made private first class at _____. I'm sorry that we can't give you more information, but that's the last we heard from him. |
| Under the aforementioned circumstances, therefore, I regret exceedingly that it will not be possible for me to send you a photograph suitable for display purposes showing what Mr. _____ is doing in our company, him having been drafted into the Army, I regret to have to say. | Sincerely, |
| Sincerely, | |

　　有些人習慣在信中寫些不必要的訊息，像左欄的原信那樣，一再重複他們認為該講的事。其實不用把每封信都寫成法律文件，前信中已討論過的項目也不用逐條重複，像右欄這樣把信簡化，讓人更容易了解：

| 原信 | 修訂後 |
|---|---|
| Dear Mr. _____ : <br><br> On _____ you wrote us concerning the price of _____ booklets you resold to _____ . You said the company claimed they could buy them more cheaply by dealing directly with us. <br><br> We replied that that assumption was correct. Because Imperial Plastic buys booklets regularly for employee reading racks they are entitled to a special price. This price, however, is obtainable only through a direct purchase. It is not available through any intermediaries. <br><br> After explaining this to you, we agreed to bill Imperial directly at the reading rack price. They have now paid that invoice. We are, therefore, enclosing our check for $_____ , refunding the money which you had already paid for these booklets. <br><br> Sincerely, | Dear Mr. _____ : <br><br> Mr. _____ has paid us directly for 1,000 booklets billed to you on our invoice No. _____ .As agreed, we are enclosing a check for $_____ refunding your money. <br><br> We're very sorry this misunderstanding arose. <br><br> Sincerely, |

## 三、刪除累贅的字與詞

說得多不見得更有效，敘事長篇累牘並非好習慣，比較一下左右兩欄，學著：

1. 消除不必要的想法、用字和用詞。

2. 把重要的部分用比較簡單的文字和比較流暢的邏輯表述出來。

| 原句 | 修訂後 |
|---|---|
| 1. Your prompt attention to this matter is urgently requested in order to protect the credit rating you had at the time this account was opened. | 1. Your prompt attention is necessary, Mr. _____, to protect your good credit rating |
| 2. It will be greatly appreciated if you will forward to us the current address on the above subject, and whether or not he is still employed by you. | 2. Do you still employ Mr. _____, and can you give us his present address? |
| 3. We are indeed very pleased to be in a position to advise you that the _____ Company is our distributor in your area and they will be only too happy to assist you on any insulating problems which you might encounter and also furnish you with a quotation on any specific list of materials or quote on an installed job, whichever the case may be. | 3. The _____ Company distributes our products in your area, Mr. _____. They will be glad to help you on any insulation problem and quote prices either on the materials or on a complete job. |

　　使用形容詞（adjective）或副詞（adverb）旨在輔助讀者了解文意，但過度使用形容詞或副詞則會增加閱讀負擔，不見得能達到真正的目的。

　　例如以下原信第一行中的副詞 *probably* 就是個贅字，就算沒有了它，這句話也已經包括了 *probably* 的內涵。再看該信第二段最後一句的形容詞 *major* 也是個贅字，就算沒有這個字，該句也已經清楚表達了 *major* 所要表達的目的了。

　　因此，在改信的時候，把文中的形容詞和副詞挑出來，只要這些字對原義沒有明顯的影響，就不妨刪去，這樣不但讓文章更生動，也讓讀者能直接了解信的中心思想。

| 原信 | 修訂後 |
|---|---|
| Dear Mr. _____ : | Dear Mr. _____ , |
| You are exactly right, and I would probably have the same feeling that you experienced form our handling of your account last month.  I am pleased to know that our people called Mr. Chung to explain the embarrassment we caused you. | You are right.  I would have the same feeling that you experienced form our handling of your account last month.  Our people called Ms. _____ to explain the trouble we caused you. |
| May I assure you that we will do everything in our control to handle your account in a proper and satisfactory manner?  Our major aim is to give you the type of service you want and expect. | We will do everything in our control to handle your account in a proper and satisfactory manner.  Our aim is to give you the type of service you want and expect. |
| A duplicate deposit, representing the refund of the $_____ service charge is enclosed.  Please accept this as a measure of our desire to serve you. | A duplicate deposit, representing the refund of the $_____ service charge is enclosed. |
| Sincerely, | Sincerely, |

　　長篇大信絕不是寫作能力的表現，反而可能是思路不清的佐證，說得太多會減弱信的力量，此時只要刪掉一些不必要的字詞，就足以把信大為改善。左欄這封信就充分顯出寫信者沒有講重點的能力，不如簡單點，像右欄那樣跟對方說清楚、講明白。讀者不妨比較原信和修改後的信，看看有何感想。

| 原信 | 修訂後 |
|---|---|
| Dear Ms. _____ : | Thank you, Mrs. _____ , |
| It is with sincere regret we find it necessary to advise you that, at the moment, we cannot accept your subscription. | for your subscription to _____ , a specialized publication edited for business executives. We've done this in order to make the magazine attractive to advertisers who want to reach the executive market. |
| As publishers of business publications addressed to specialized audiences, our circulation rolls are in a very real sense restricted as compared with the mass circulation interests of any general magazine. One of the serious problems to the publisher of a business magazine is how to gracefully postpone the acceptance of subscriptions at a time when certain circulation classifications are at capacity. | To keep these advertisers interested, we have to restrict our non-executive readership. Presently, we have all the non-executive subscribers we can safely carry. |
| We are therefore making arrangements to refund your subscription fee. However, our circulation requirements do vary and it is quite possible that, if you would contact us at a later date, we might be in a position to accept your subscription to our publication. | Sorry to disappoint you, Mrs. _____ , but I'm sure you'll understand our predicament. We have to respect the needs of our advertisers or go hungry. Perhaps you can persuade some executive friend to share his copy with you. I surely hope so. |
| Thank you very much for your kind consideration. | Sincerely, |
| Sincerely, | |

　　以下左欄所列的是些贅字贅詞充斥的句子，贅字詞把許多該表達的主旨都淹沒了，不妨把那 14 個句子簡化成像右欄那樣：

| 原句 | 修訂後 |
|---|---|
| 1. Please feel free to write to us if you find yourself in need of more information. | 1. Please write if you need more information. |
| 2. If at any time whatsoever we can be of additional service to you, kindly to not hesitate to let us know. | 2. If we can be of additional service to you, let us know. |
| 3. We want to thank you very kindly again for your interest in Blank products. | 3. Thank you again for your interest in Blank products. |
| 4. This is to express my personal thanks for your order. | 4. Thank you for your order. |
| 5. Per your request, you will find enclosed is information on Blank valves. | 5. Enclosed is the information you requested on _____. |
| 6. May we ask that you kindly permit us to place your name on our mailing list? | 6. May we place your name on our mailing list? |
| 7. In accordance with your recent inquiry at hand, we are pleased to be able to send you herewith a copy of our very latest catalog. | 7. We are pleased to be able to send you a copy of our latest catalog. |
| 8. This will acknowledge with thanks your letter of March 1, 1960, which is very much appreciated. | 8. Your letter of March 1 is very much appreciated. |
| 9. We are anticipating being in the position of shipping your order soon. | 9. We expect to your order soon. |
| 10. The undersigned wishes to acknowledge with deepest thanks receipt of your kind inquiry. | 10. Thank you for your inquiry. |
| 11. Complying with your request of recent date, you will find attached is a copy of our very latest price list for your perusal. | 11. Attached is a copy of our latest price list. |
| 12. The attached self addressed, postage-free envelope is for your convenience in sending us your prompt reply. | 12. The attached, postage-free envelope is for your reply. |

| 原句 | 修訂後 |
| --- | --- |
| 13. Whenever we can be of help to you, kindly do not hesitate to let us know. | 13. Whenever we can be of help let us know. |
| 14. May we take this occasion to thank you kindly for your interest in our company? | 14. Thank you for your interest in our company. |

　　不妨檢視一下自己的文章，看看有沒有同樣的毛病，如果有，不妨大刀闊斧地整頓一下，表 4-1 是一些累贅的寫法和簡潔寫法的對照。另外，贅詞簡化表所列的是些贅字詞及其清晰有力的代用字，請將兩者互相比較，盡量捨繁取簡。

表 4-1　化繁為簡

| 累贅的寫法 | 簡潔的寫法 |
| --- | --- |
| at the present time | now |
| at your earliest convenience | soon（或 promptly, immediately） |
| I have insufficient knowledge | I don't know...（我不知道） |
| In my opinion it is not an un-justifiable assumption | I think...（我認為……） |
| in the very near foreseeable future | soon（或 promptly） |
| It has long been known that(早已為人所知) | I believe...（我相信……） |
| It is believed that / It is suggested that | I think...（我認為……） |
| It is my understanding that | I understand...（我了解……） |
| It may be that | I think...（我認為……） |
| It was observed in the course of the experiments | I observed...（我觀察到……） |
| much additional work will be required before understanding | I don't understand...（我不了解……） |

| 累贅的寫法 | 簡潔的寫法 |
|---|---|
| My attention has been called to the fact that | Later I found...（我事後發現） |
| The opinion is advanced that | I think...（我認為……） |
| There is reason to believe | I think...（我認為……） |
| under separate cover | separately...（分別寄出……） |
| We are this day in receipt of | today we receive...（今天收到……） |
| We wish to thank | We thank...（我們感謝……） |
| Your check in the amount of $15 | your $15 check...（你的 $15 支票……） |

## 四、直截了當恰到好處

好的 e-mail 開頭宜簡要、容易了解，又能引起讀者的興趣。

左欄這個例子是篇文章的開頭第一段，這段文字若是演講稿的開場白也就無可厚非，但若用它作賣房子的銷售信的首段，那就顯得在賣弄風格了，因此不妨把它簡化如右欄。

| 原文 | 修訂後 |
|---|---|
| From early puritan household to weatherworn pioneer family, from Black/Irish/Chinese immigrant household to Apache/Sioux/Navaho family, from early log cabin to modern suburbia, the strength of America has always been—and still is—the American family. | From early log cabin to modern suburbia, the strength of America has always been—and still is—the American family. |

同上理，你可以簡化左欄的文字，讓它像右欄那樣，更切合要點。

| 原文 | 修訂後 |
|------|--------|
| Along with the objective criteria of taxes, transportation, utilities, and so on which we use in choosing a new plant site, we try to get a feel for the economic climate of a community under consideration. | In choosing a new plant site we try to get a feel for the economic climate of the community under consideration. |

　　以下是一段切中要點的開信詞，沒有暖場的贅語，不賣弄文字，非常合乎讀者的心理：

From one builder to another...

Let me tell you about a merchandising program that works. I know because we used it here in Peoria to successfully promote our new willow knolls development.

　　親切的短信可以不用一般的客套開頭語（如某某先生鈞鑒之類的）。請輕鬆想像受信者就坐在你身邊，你在用與人交談的字眼來寫這封信。

　　簡短的信無疑是比較合適的，不用擔心信會太短，而要把重點放在答覆對方問題上，並用愉快的方式結尾。

　　信的內容多寡本來就沒有一定的原則，只要足夠表達寫信者的意思也就好了。

　　話說回來，簡短的文章雖然容易了解，但若沒能摘出重點，用途也是有限。請看左欄這篇回答顧客洽詢的例子，這封信簡短有禮而且中肯扼要，將問題回答得令人滿意，但它的缺點在推銷功能不顯，因為它並沒有嘗試增加對方的興趣，或者引發對方下單的念頭。看看右

欄那封改良後的信，特別注意其中斜體字部分的推銷字句。

把銷售的感覺放進信裡，以便吸引有意消費的人前來詢問。

| 原信 | 修訂後 |
|---|---|
| Dear Mr. _____ : | Dear Mr. _____ : |
| Thank you for your interest in _____ . The cost of _____ copies is $_____ each. It would be a pleasure to serve you. | Thank you for your interest in _____ . *It is one of our best sellers. Many companies use it in their training programs.* |
| | The cost _____ would be $_____ per _____ copies. It would be a pleasure to serve you. |
| Yours truly, | Yours truly, |

## 避免措辭公式化

　　早期的英文書信傾向於使用制式（即樣板式）的措辭，久而久之，大家沿用成習，若不照著樣本寫信，反而被人引以為怪。有些管理階層的人喜歡用些特別的文句來表達精確的文義，這種方式用在法律性的文件（如合同及條約等）上，或有必要，但若用在給員工、客戶及大眾的資訊上則大可不必。

　　舊式的公務用詞有時仍會出現在現代的書信裡，這些用詞語意纏夾，少了它們無損於信的原意。以前的人注重格式與身段，甩不掉那些陳詞老調，殊不知，寫信就要簡單直接，那些舊語言早就失去了時代意義，應該用現代白話文來取代，信越白話易懂，越表示寫信者的表達能力高人一等。

　　在傳達任何事情的時候，首先要考慮的就是傳達的對象，既然寫的不是法律條文或商業合同，那又為什麼要疊床架屋，重重修飾，以

致要人大費周章才能明白文中的意思呢？

一、揚棄陳舊文字

　　下文是用英文書信的刻板措辭拼湊出的順口溜[2]，幽默而帶有取笑的意味，可以博君一笑：

We beg to advise you, and wish to **state**,
　　that yours has arrived of recent date.

We have it before us, its contents *noted*;

Herewith enclosed are the prices *quoted*.

Attached you will find, as per your request,
　　the sample you wanted; and we would suggest.

Regarding the matter and due to the *fact*
　　that up to this moment your order we've *lacked*,
　　　we hope you will not delay it *unduly*,
　　　and beg to remain yours very *truly*.

　　下表左欄的信包含了長度相近的 5 個段落，其中四段是以 *If* 開頭，信件的內容太過於刻板，像是套自書信大全。

　　簡單明白一定比拐彎抹角更讓人容易接受，明智的人會用讓人容易明瞭的白話口語來寫信，不明智的人才用非常饒舌的字句去衍繹他的信。因此，不如將原信以右欄那樣的方式重寫：

2　請注意用黑體、斜體或底線標出之每兩句尾押一韻的部分。

| 原信 | 修訂後 |
|---|---|
| Dear Friends, | Teachers are busy people, Ms. _____ |
| We are enclosing a bill covering examination copies of some of our texts recently sent to you on approval. | but I do hope that you've had a chance to review the examination copies of the texts we recently sent you. |
| If, after an examination period of thirty days, you wish to return any or all of the books, you may do so. | An invoice covering the books is enclosed. You don't have to pay this invoice unless you decide to keep them. Otherwise, just return the books within thirty days. |
| If copies are ordered for classroom use, in the amount of twelve or more, desk copies of the texts purchased may be retained without payment. | If you buy twelve or more copies of any one of these titles, you may keep the examination copy free of charge for desk use. |
| If the examination copies are kept for personal use, or library reference, a remittance may be made at net price at the conclusion of the thirty-day examination period. | I hope you'll like these texts, Mr. _____. In our own prejudiced way, we think they're excellent. If you'd like further information about any of them, please let me know. |
| If you wish further information on any or all the books sent, please let us know. We shall be happy to be of service. | Sincerely, |
| Sincerely, | |

　　用制式的文字寫信只會顯得古板，以下左欄這封信就是個古板的例子，像右欄那樣交談的通信方式可用更好、更短、更愉快的文字完成同樣的事。

| 原信 | 修訂後 |
|---|---|
| Gentlemen[3]: | Dear Mr. _____ : |
| In accordance with your recent postal inquiry we take pleasure in sending herewith a copy of our latest catalog herein our line of Evaporative Condensers and Cooling Towers in detail. | It is a pleasure to send you a copy of our latest catalog describing _____ . Here, too, is a copy of our Installation, Operation, and Maintenance Manual. |
| We also take pleasure in sending you the Installation, Operation, and Maintenance Manual of our Evaporative Condenser which we trust will prove to be of interest to you. | Thanks for your inquiry, Mr. _____ . If you have any questions about these products, please let us hear from you. We would be happy to explain the features, which make it an outstanding value. |
| If you should have any questions regarding our products, kindly do not hesitate to communicate with the undersigned. | Yours truly, |
| Thanking you for your interest. | |

## 二、不要官腔官調

　　公司行號中一些與法律有關的信件（例如催欠信等），多會由學法律的人來撰寫，他們之中有許多人受過專業訓練，精通法律卻忽略了單純的人情，只管寫些冷硬的文字。

　　從事法律業務的人也該學習寫些溫馨感人的信件。

　　且看左欄這封樣板催欠信，改成像右欄那樣友善也完全表達了催討的目的：

---

3　如同 Dear Sir, Dear Madam 等這類沒有指出特定收信者的稱呼，只宜在不知對方姓名稱謂時使用，因為其禮貌性永遠比不上稱呼受信者大名的方式。

| 原信 | 修訂後 |
|---|---|
| Dear Mr. _____ : | Dear Mr. _____ : |
| You are hereby notified that pursuant to the terms of your installment contract No. IL-18-007 for a total of Seven Hundred Nine and 06/100 ($7006), you are in default in the payment of the following payments: | According to the terms of your installment contract, payments of $____ were due on __/__/__ and __/__/_____, but they have not arrived. |
| 1. November 16 $_____<br>2. December 16 $_____ | As you know, there is a clause in your contract, which says that under such circumstances the bank may declare the entire unpaid balance due and payable. As much as we have tried to avoid it, this action now becomes necessary in order to protect our depositors and stockholders. |
| and that by reason thereof the undersigned bank has exercised its option contained in the above-mentioned instrument to declare the entire unpaid principal now due and payable, and in compliance with Federal Housing Administration requirements for the filing of claims, hereby demands payment forthwith of the entire balance so due, plus late charges which items now amount to Five Hundred Nine and 48/100 ($590.48). | Accordingly, we must demand that the entire balance plus late charges, a total of $_____, be paid immediately.<br>Sincerely, |
| Sincerely, | |

　　下例左欄的信是封不好的示範，寫信者或許是個好人，可是他似乎在認知上有錯誤，以為公務書信就該這樣寫。右欄這封信與原信意思相同，但用了不同的表達方法，比較看看，你願和寫哪封信的人做生意？

| 原信 | 修訂後 |
|---|---|
| Gentlemen: | Dear Mr. _____ : |
| Pursuant to yours of the __/__/_____ , which valued communication has been brought to the writer's attention, the writer is cognizant of the urgency of the situation relative to the delay in shipping the _____ parts. However, under the prevailing circumstances and due to the fact that there has been an unfortunate and unavoidable delay in obtaining the requisite steel, it will, in all probability, require the expenditure of an additional amount of time in the amount of seven days, more or less, before we expect to be in a position to initiate shipment. Trusting that this will explain satisfactorily our sincere and earnest desire to be of whatever assistance we may under the aforementioned circumstances.<br><br>Sincerely, | You may be sure that we, too, are disturbed at our inability to ship you the parts you need so badly for your Model _____ . It may not comfort you much to have me say that the steel shortage is the sole villain this time—but it's the truth.<br><br>There is one ray of hope, however. I think I can ship you these parts within a week, but be shipped as soon as possible, Mr. _____ .<br><br>Sincerely, |

以下左欄這封也是標準的樣板信，其中想表達的訊息可用像右欄那樣簡潔的方式來表達。

| 原信 | 修訂後 |
|---|---|
| Dear Sir: | Dear Mr. _____ : |
| Re. Yrs. of the 5th inst. at hand, are mailing under separate cover bulletins and literature covering our latest machines. Thank you for your kind inquiry and hoping to be of service to you in the very near future. | Thank you for your letter asking for information on our latest machines. It is a pleasure to send you some of our bulletins and literature. If you need more, please let us know. |
| Very truly yours, | Yours truly, |

　　看看下面左欄的舊式短簡，將它改成右欄那樣後，兩者語氣大爲不同。原信中的 *We wish to thank you for...*，被 *Thank you for...* 取代，而 *We shall strive to merit your confidence* 爲 *We will do our best to serve you well* 所取代。

| 原信 | 修訂後 |
|---|---|
| Dear Mr. _____ : | Dear Mr. _____ : |
| We wish to take this opportunity to thank you for your commercial account opened at this office today we shall strive to merit your confidence in the years to come. | Thank you for opening a commercial account at this office today we are glad to have you with us and will do our best to serve you well, Mr. _____ . |
| Sincerely, | Sincerely, |

　　重要的書信或報告應好好斟酌，也不妨請別人幫忙校讀，看看人家的意見如何，看看人家是能否輕易看出你的意思。

## 三、變換表達方式

　　字句的經營很重要，好好推敲信中的文句，相似的意思可以用不同的說法來表達，用不同的句子來表達可能更吸引顧客，請看下列的例句：

1. The curtain is easily torn.  It is made of flimsy material.
2. The curtain is easily torn, for it is made of flimsy material.
3. The curtain is easily torn because it is made of flimsy material.

4. The curtain is made of flimsy material and was easily torn.

5. The curtain is made of flimsy material which was easily torn.

6. The seat cover, which is made of flimsy material, was easily torn.

7. Being made of flimsy material, the curtain is easily torn.

8. The curtain is made of easily torn, flimsy material.

9. The flimsy curtain is made of easily torn material.

10. The flimsy curtain is easily torn.

　　要加強寫作能力就得付出努力，別無他法。你的信要更宜人、更得體，可以參照下面這幾個原則：

1. **刪除不相關的陳述**：有人寫信時會來段不觸及主題的開場白，這類開場白在演講時有暖場效果，但在講求簡潔的信裡就沒有必要，不妨刪掉這類不相關的文字。

2. **刪除不順暢的陳述**：把完成的信朗誦一遍，在朗誦過程中若碰到唸起來不很通順處，就該刪去或重寫，至於那些突兀而與上下文意不太相容處，當然可以大刀闊斧地除掉，這樣可使整封信的文路更暢通，使意思表達得更平順。

3. **不要咬文嚼字**：有些賣弄才情的人愛在文中加些自以為傲的「優美」文句，以為能增加信的文藝氣質，但其實這類文字的實用性有限，不宜出現在像報告和公務書信等應用文上，因此也在刪除之列。

4. **事件的順序宜合乎邏輯**：把事件依重要性從低到高排列出來可以增加其可讀性，也可以讓讀者比較容易抓住重點。

5. **要認同讀者**：認清讀者群，採取他們較易接受的字詞和用

語，迎合他們的風格。

 ## 訴求宜明顯統一

信裡的主題越多，就越難把每個主題都表達清楚，成功的書信應有磁鐵般的主題，能把信中必要的事情吸附在一起，信的內容應該圍繞著該主題，其他無關的事不妨排除。

請集中火力，一封信只銷售一種產品或服務，不要想同時賣抽水馬桶和微波爐。

請看左欄這封信，一開始想建立輪胎專家的身分，但隨即又轉而介紹電池和其他修車服務，這雖不算什麼大錯，但是何不先立穩了輪胎的招牌，之後再在另封信裡提出別的項目？請見右欄。

| 原信 | 修訂後 |
|---|---|
| If you have a tire give out anywhere within five miles of _____, call _____. We have full conveniences, a complete store of tires, tubes and accessories. | Do you have tired tires, Mr. _____? If so, you know that sooner or later you'll have to do something about them. Why not drive in and get the free advice from our experts? |
| We have established as a service station for _____ tires. We chose _____ tires after several years' experience because they proved to be the tires that stand up best on our customers' cars. | In tires, as in most everything else, it pays to consult a specialist. Here at _____ we specialize in complete tire service such as periodic inspections, repairs, recapping and retreading, and new tires as well. |
| When your tires need care, just drive up in front and blew your horn. Our men are always at your service. | Just drive up and honk your horn. One of our men will look over your tires and tell you exactly what he recommends and what it will cost. There's no charge for his advice—you can take it or leave it. At least you'll know the fact—that's a lot smarter than just driving |

| 原信 | 修訂後 |
|---|---|
| Our garage is the cleanest and most inviting place to leave your car you'll find in the country. It's fireproof, and roomy enough to turn in without danger of bumping.<br><br>We maintain a service station for ___ ____ batteries and are equipped to do all kinds of battery overhauling. Our repairmen are experts in all kinds of automobile repairs. You'll find them just as familiar with your car as a doctor is with the ailments of your family.<br><br>Take advantage of our service. It will mean a saving to you. | along until you have a tire failure, perhaps a serious one.<br><br>You may not need new tires. Recapping or retreading may give you thousands of miles of safe driving at minimum cost. If you do need new ones, we carry a complete stock of _____ Tires. Experience with thousands of customers has convinced us they are the best on the market.<br><br>Tired tires are dangerous to ignore. Why not stop by and let us look at them for you?<br><br>Remember if you ever have tire trouble within ten miles of _____, just call _____. We'll be there pronto! |

　　不論信中的主題為何，其他所有的條件都應為主題服務，如若不然，不妨刪除。若有不只一個主題，最好分成幾封信來寫，免得主題散亂、宗旨不明。

Volume

# 2

# 公務書信體例

本篇提要

# 第五章
# 銷售信的藝術

本章提綱

◎銷售信的寫作
　寫作原則
　寫作方法
◎銷售信的心理
　不用喜新厭舊
　避免過於強勢
　揣摩顧客想法
　採取後續行動
◎招攬新客戶之道
◎贏回舊客戶之道

　　一般人收到紙本或是 e-mail 銷售信，只會大略瞄一下或直接丟到垃圾桶，少有人花時間或一一讀完。因此，如果要引起較多的注意，就需引起讀者濃厚的閱讀興趣。

　　什麼樣的東西會引起讀者的興趣呢？

　　人總是比較關心與自己有關的事，你的外表、地位、別人對你的看法……，你都會投入相當的關注，因此，報上才會有這麼多星座分析、相命等文章。

　　你在閱讀生肖或星座運勢的時候，多半先讀有關於自己的生肖和星座。一封信的主題如果與讀者有切身關係，那麼你只要一開始就端

出主題，讀者就容易有興趣讀下去。

　　因此，若想寫封廣受閱讀的文章，請繞著讀者的興趣打轉。

 銷售信的寫作

　　在某種程度上，每封信都是銷售信，我們用它來讓人有好印象，進而加強其購買的意願。

　　書面銷售的重心是讓人心動而不是讓人心煩，說動客戶下單的重點在於令他心動的誘因，製造不出誘因的信乃是無效的信。

　　好的促銷策略在於利用一封短信來引起閱讀慾望，再伴隨附件來加強銷售的內容。一開始若寫些無關緊要的事，又夾帶許多宣傳單，會減低閱讀的興趣。

## 一、寫作原則

　　銷售信應該有多長？

　　銷售信若太短，也許無法達到預期的效果，若太長，就難以期望人家不嫌麻煩地讀完，自然也就難達預期目標。因此，促銷信應該只要長得足夠達成目標。

　　下列是銷售信的幾個重要原則：

1. 有明確的目的：若旨在吸引進一步探詢，就要讓讀者對你的東西產生足夠的興趣。目的如果在取得訂單，就得盡量回答顧客考慮下單時所提出的問題，並且提供充足的資訊以求提高其購買的可能性。

2. 列舉優點：告訴客戶你的產品或服務對他們有何好處，把那些好處按照重要性列舉出來，最重要的排在最前面。

3. 消除客戶可能有的疑慮：對於買方疑慮，你若有把握消除則

不妨一一列出，各個擊破。若有不能自圓其說的缺失，則大可不用自曝其短，必要時用加強優點的方式將它帶過。

4. 找出最大的賣點：把商品或服務的最佳賣點列在標題上，當它是新聞中的頭條，把信的主旨繞著那個賣點轉，用商品的其他優點來烘托這個賣點，並找出實例來支持這個賣點。注意，用詞要中肯可信，千要不要太誇張而導致反效果！

5. 邀請顧客訂購：在請對方下訂單的關鍵時刻，你應該以顧客的立場來著想。想想看，自己若是顧客，會在什麼樣的誘因下購買？是免費試用呢？還是不合用保證退費？或是無須立即付款？額外的贈品又如何？這些誘因的效果因人而異，值得多花腦筋，最主要的是要以顧客的方便為考量的重點。

6. 使用便利的回覆卡：一切以客戶方便為考量，要請客戶回覆的信件應附上包含郵資的回郵卡，使收信者能隨手覆信，加速商機。提供免費服務電話來強調售後服務，也可穩定客源。

7. 避免過於繁瑣：好的業務書信裡敘述的主題不宜過多，以免混淆，最好是一信一主題，尤其 e-mail 發達，把不同的主題分別寫在不同的 e-mail 裡分開寄發，並不會多花郵費。

8. 慎用附記（P.S.）：公務信件裡宜避免使用 P.S.，因為 P.S. 有時會讓人覺得寫信者太大意，忽略掉了重要的事件，到了信末想起來才補加個 P.S.，不太可取。但經過刻意經營的 P.S. 也可能發生很好的效果，尤其對大宗的銷售信，收信人瀏覽此類信件的速度通常很快，不易抓住重點，此時信尾的 P.S. 可以重新抓回讀者的注意，只要寫得有技巧，讀者可能會因而回頭把信再讀一遍。

二、寫作方法

　　根據上述的原則，設定中心主旨，然後把要表達的內容依照下述的步驟一一道出。

　　**1.** 引起注意：先用有力的開頭來引起讀者的共鳴。

　　**2.** 引起興趣：其次要告訴讀者，信的主題跟他們有切身的關係。

　　**3.** 引起慾望：然後使讀者有不妨試試的念頭。

　　**4.** 引起信心：下一步要讓讀者相信，你鼓勵他們做的事錯不了。

　　**5.** 引起行動：說服讀者心動不如行動，並要幫忙去除妨害行動的障礙。

　　下面這封信就是個很好的範例，它不但足以引起讀者的注意和興趣，也有個足以引起行動的結尾，我們把它分段標出如下：

　　第一段引起注意和興趣：

What are <u>you</u> looking for in ceiling air diffusers?

1. Draftless air distribution?

2. Quick temperature equalizers?

3. Low cost and pleasing appearance?

4. Simple, cheaper installation?

　　第二段引起慾望：

The FLEXIFLO diffuser offers <u>you</u> all these features <u>plus</u>:

1. <u>Adjustable after Installation.</u> Just a simple twist of the wrist to control volume

2. No Auxiliary Dampers Needed!

3. Built-in Equalizing Deflectors on every unit.

4. Quick Deliveries. Many sizes stocked.

5. A Full Line. All type of diffusers and accessories.

第三段引起信心：

We are able to offer these engineering features at competitive prices because of our unique patented construction with drastically cuts manufacturing costs.

FLEXIFLO is universally accepted as the greatest advance in diffuser design.

第四段引起行動：

Please let us send you our Engineering Manual. Your can get this book, including many useful facts, by calling our representative in principal cities, or by mailing the enclosed postpaid reply card. No obligation to you, of course.

　　要引起顧客的興趣，就要使他們相信你的資訊可靠而重要。只要能清楚提供客戶所期望的資訊，你的信就已經成功一半，客戶一旦被說服，就會有反應。當然，爲了客戶方便，就應該提供回郵卡。

　　要靠銷售信來吸引訂單，除了上項引起興趣的原則外，還要考慮

到客戶：

　　1. 需要什麼進一步的資訊。

　　2. 需要什麼樣的證明。

　　3. 需要下什麼樣的決心才會下訂單。

　　你若能引導客戶做好這類的決策，接下的事就好辦了。

　　也許你擔心有人不想閱讀長信，但若因此而把信縮短得失去說服力，也是矯枉過正。因此，不妨暫時別管那些嫌麻煩不讀信的人，先指向那些可以從你產品獲益的人，提供他們所需要的資訊，討論他們切身的問題，提供證據，證明你的東西確能滿足他們的需要。

　　推銷信如果很長，不妨在訂購卡上擇要敘述信上的賣點，訂購卡有時比銷售信還重要，有些人會被訂購卡引起興趣，直接下訂。因此，設計有活力和說服力的訂購卡是很有用的。

　　參考一下雜誌或別家的文宣，看看是否會引起你下單的興趣？如果是，你不妨拿來效法。事實上，這些文宣都是通過檢驗的，大公司花了許多廣告費才發展出這些富有推銷力的作品，你大可好好利用。

##  銷售信的心理

　　能用簡短的信銷售成功當然很好，但切莫光顧著要簡潔而省略了基本原則。

### 一、不用喜新厭舊

　　成功的銷售信值得一用再用，即使你確信有封新的信件會產生更好的結果，也不用拋棄這封有過銷售成果的舊信，除非你證實新的信件有更好的促銷成績。

　　寄了信給某通訊錄裡所有的人後，若發現大有斬獲，那就表示這

封信的確有效，應該還有很多人對你的產品有興趣，也許有人原本想下單，卻因為雜事延誤而遺忘。那麼，不妨幾個月後把這信再寄給同樣的人。

第二次的成果可能不如第一次好，但根據經驗，效果可能達到第一次的 60%。

二、避免過於強勢

書面推銷與當面推銷有很大的不同，直銷中有一種所謂的「高壓銷售法」，用鍥而不捨的纏功造成對方破財打發或不好意思不買的心理而銷售成功。這種方式在郵購業務中絲毫無用武之地，客戶只要一覺得沒興趣，那你辛苦寫的信就進了字紙簍。

左欄這封信是某公司的業務向客戶推銷自己公司服務的信，信中口氣逼人，並且自作主張地訂下見面的時間及地點，甚為強人所難。

要引起客戶的興趣，只能用「吸引」而不能用「強銷」，對方受逼的反應多半是反彈，所以不要強求對方照你的意思。右欄是經過修改的例子，寫信說服別人時，要把自己當作是收信人，自問是否能接受信中的說法。

| 原信 | 修訂後 |
| --- | --- |
| Dear Mr. _____ : | It was a real pleasure, Mr. _____ , |
| I visited you about July 17th. The subject was our Research Method of Business Development, in which you expressed interest. | to discuss our Research Method of Business Development with you. You were so quick to appreciate its potentialities for your bank that there seems little more I can say. |
| The intent was that you would arrange for a meeting about this time. | |

| 原信 | 修訂後 |
| --- | --- |
| The purpose would be to work up every detail of schedule, forms, procedure and plans for a starting date. <br><br> May I suggest that we meet on September 10th. I will plan to be in Tulsa for the purpose. <br><br> If however, this date is inconvenient, be good enough to let me know, suggesting a more convenient date. Not hearing to the contrary, please look for me on the 10th. <br><br> Sincerely, | You suggested that the next step would be a meeting to work out details of a possible schedule, forms, procedures and plans for a starting date. If it is convenient for you, I could be in Tulsa September 10 for such a meeting. If you'd prefer another date, just let me know, and I'll do my best to make it. <br><br> It's an excellent time to inaugurate a program of this kind. If you're ready to investigate further, we'd be delighted to work with you. <br><br> Sincerely, |

　　在習慣於抗爭或大聲疾呼的環境裡，柔性訴求可以造成一種截然不同的清流印象，效果不見得遜色，大家不妨試試像下面這封信一樣，用溫柔的方法傳達訊息：

The same man, with the same eyes, can see things the wrong way. All he has to do is stand on his head.

The jeweler with both feet on the ground tries to make every sale. He tries with the best selection possible, and the best values in the market.

We can and we will give you both.

下面是另一封成功的柔性訴求的信：

A lady, suing for a divorce, stated that her husband was very careless about his appearance. He hadn't shown up in almost two years.

Not so with us. Just give us the least bit of encouragement, and our salesman will show up right away.

三、揣摩顧客想法

信的優劣關鍵，在於能否將意思清楚表達給收信人，所以你應該揣摩收信人的想法。左欄這封是由泳池包商寫給建商的信，寫得雖然還可以，但若能揣摩一下收信人的立場，還會更好。

想想看，該建商在什麼情況下會對這包商感興趣？

1. 這家要價比較低。

2. 這家的施工品質比較好。

3. 不滿意目前的合作對象，有意另找一家試試看。

基於這 3 點可能的原因，原信可以重寫如右欄：

| 原信 | 修訂後 |
|---|---|
| Gentlemen: | Your name, Gentlemen, |
| I am sending you this brief, get-acquainted note as a result of my reviewing the current issues of the "Builders' Weekly". George Wick, Inc. has been and shall continue to be extensively and progressively | has come to my attention several times lately, most recently in the current issue of the _____ . As developers, you have at least occasional need for swimming pool installations. I'm sure |

| 原信 | 修訂後 |
|---|---|
| interested in the design and construction of superior quality swimming pools and related facilities.<br><br>We would be extremely appreciative of being placed on your bidders' list and placed in your active files for any services that may be required for the installation of a complete swimming facility.<br><br>I am enclosing one of our brochures which includes a partial list of commercial swimming pool projects that we have had the good fortune of constructing. If there are any projects requiring immediate bidding or any other services, please call me and I will assure you of my immediate attention to whatever the need may be. If there is nothing of an immediate nature, please call us as the need arises.<br><br>Sincerely, | I needn't tell you how widely swimming pools can vary in quality, how much damage an inferior installation can do to a developer's reputation, and how expensive subsequent repairs can be.<br><br>If you have complete and utter confidence in your present subcontractor, perhaps there's nothing we can offer. If you don't, I'm sure it would be of mutual benefit for us to sit down and have a talk.<br><br>I will phone you shortly to discuss this matter further. In the meantime, if you have any questions or projects which require immediate bidding, I would be most happy to receive your call.<br><br>Sincerely, |

　　許多銷售失敗的原因，在於沒設身處地為顧客思考。郵購或網路銷售成功的祕訣跟任何其他生意都相同，就是要與顧客站在同一個角度去看待事物。

　　顧客喜歡的賣點是：

1. 交易方便——沒有人喜歡花錢找麻煩，因此要設計些簡單的交易方式，例如信用卡簽帳、快遞送貨等。要請客戶回覆的信件不妨附上包含郵資回覆卡，提供免費洽詢電話也很有

用。總之一切以客戶方便為考量。

2. **沒有風險**──沒有人喜歡受騙，因此可用免費試用，無效或不合用時保證退費，免先付款等方式來吸引顧客下單。

當然還有其他誘因可以促銷，而總不外乎投其所好這個大原則。但投其所好也有個限度，以左欄這段銷售信的開頭為例，這種浮而不實的開信語好像在推銷橫財夢，能有多少效果？連你自己都不信的事，要怎麼期望客戶去相信？

描述事實避免浮誇，才能夠厚築誠實的形象，對日後的商譽有強大的助益，因此不如把它改成右欄這個樣子：

| 原信 | 修訂後 |
| --- | --- |
| Dear Builders: | Most builders, Mr. _____, |
| The few minutes you spend going over the enclosed literature is apt to prove the most profitable time you've spent in a long time. | are in business to make money, not for their health. The enclosed brochure shows how our franchised owners are making money. |
| …… | …… |

若你有專業背景，在寫信時該特別注意，有些你能完全了解的事，別人可能覺得很難懂，你工作上有些常用的字彙也許令別人很費解。因此，請設身處地想一想，銷售信應該避免太過專業化。

## 四、採取後續行動

發信後有時還要有後續行動（follow-up），像打電話，親自拜訪，甚或兩者都做。因為該信也許足以開啟商機，但不見得可以幫忙拿到訂單，因此後續的接觸就很重要。

不論求職或求售，積極的態度都有必要，好的求職或求售信可以有助於打開一扇門，但是其後的追蹤洽詢也不該忽略。

有人會想採取某些行動，但事到臨頭卻又忘了，例如：有些客戶可能在收到你的信後，準備找時間好好研究或下訂，但是因為某種原因而忘了這回事，等到想起來，那封信已經不見了。在這樣的情況下，你若有寄隨訪信（Follow-up Letter）的習慣，就有機會再啓商機。

 招攬新客戶之道

為了招攬客戶，某汽車廠寫了左欄這封信，其中缺乏感情的商業術語過多，而且一直在幾個主旨間遊走，意思不清，好些想法沒能表達出來，因此沒有得到預期的效果。建議挑出那些毛病，改寫成右欄那樣：

| 原信 | 修訂後 |
| --- | --- |
| Dear Mr. _____: | We've missed you, Mr. _____. |
| Enclosed bulletin has recently been mailed to all of our active wholesalers. Included are many new items for servicing this year's new models. | You haven't placed an order with us in fifteen months. That isn't good for us—we doubt if it's good for you. Chassis parts are big items in the service parts dollar. Other wholesalers handling our line are enjoying a good turnover and an excellent mark-up. They don't have to carry large stocks to do it, either. |
| Wholesalers handling our line witness a good turnover and gross profit with minimum stock requirement. | |
| As we have not had the pleasure of serving your firm for some time, and in the event that we have fallen down somewhere along the line, we would appreciate hearing from you. We have | What's wrong, Mr. _____? Where have we fallen down? How have we disappointed you? Won't you please take a minute of your time to drop me a note on |

certainly valued your business and would like very much to have you with us again.

Chassis parts enjoy a large percentage of the service parts dollar. It will be our pleasure to explain our current wholesaler's program if you are interested.

Very truly yours,

the enclosed card and tell me?

I'd consider it a personal favor if you would. And I'll make it my personal concern to see that whatever's wrong is straightened out to your complete satisfaction. We want you back on our team.

Sincerely,

P.S. Have you seen the enclosed bulletin? It lists a number of new items for servicing this year's models.

　　伏筆是小說家和說書人所用的有效策略，它可製造懸疑的氣氛，引起讀者繼續往下讀的興趣。好的小說靠此吸引讀者，好的信件也是如此。

　　在讀你的第一封信時，讀者對你的東西可能沒有先入的觀念與了解，因此銷售信最好製造點伏筆來引起興趣。一旦他們的興趣被引發，你就有成功的希望。因此，信的開頭別老是講自己如何如何好，而要說你的商品或服務對讀者有何不可或缺的幫助。

　　另一方面，來信洽詢者對你的商品可能已有興趣，此時你就要清楚切題地回答客戶的詢問，不宜再賣弄神祕。

 贏回舊客戶之道

　　花力氣吸引新客戶，也應熱心抓住老客戶。

　　除了顧客消費時要親切相待，對那些有段時間沒光顧的客戶也該有所表示，才算是好的行銷策略。

　　某公司發現業務下滑後，寄給那些久未上門的顧客以下這封信，寫得誠懇有禮，讓顧客感受到該公司很在乎他們的生意。

We've missed you, Mr. _____.

I notice that you haven't done business with us for some time (or that your business with us has dropped off considerably). Not having heard from you, I've been wondering if you've been dis-satisfied for any reason. If so, won't you please let me know so that we can make amends?

As a previous customer, you're a key figure in our scheme of things. Reorders from old customers are more important to us than new orders. They are the best indication that we are living up to our reputation for superior quality, honest values, and ex-cellent service.

I hope nothing's wrong, but if there is, we'd certainly appreciate an opportunity to make it right. Won't you please drop me a note on the enclosed card?

　　生意清淡時不妨想些點子來吸引顧客，例如，給久未上門的顧客寄張問候卡或提供小禮物，這種百貨公司促銷的方法也可以運用在其他類型的生意經營上。

　　要贏回以前的客戶，就得寫出能讓他們回心轉意的內容，下面這封信是很好的參考：

Dear Mrs. _____:

Like all good housekeepers, we like to tidy up an extra bit for the holidays. So, with the _____ Shopping season at hand, we discovered that your credit card hasn't been used in months. "This," we said to ourselves, "is something we'd better look into right now!"

Have you lost your credit card? Do we have your correct name and address?

Your credit card is a friend indeed when it comes to shopping, and we know you'll want to use it especially for _____. Our exciting new fashions are here for you. Come in and see them soon, for it's a good idea to shop early for the largest selection.

Will you please give us a minute of your time now and answer the questions on the enclosed postcard? It requires no postage, and if you'll speed it along to us, you'll end our quandary about your credit card, Mrs. _____.

Cordially yours,

# 第六章
# 深得人心的回信藝術

本章提綱

　　◎不要怕說對不起
　　◎善解顧客的抱怨
　　◎避免敷衍的態度
　　◎注意回信的情緒
　　◎婉拒的藝術
　　　　婉轉的重要
　　　　客戶的信用

　　迅速回信有助於掌握商機。

　　今天就把事情做好和拖到下個星期再做，要花的工夫都少不了，何不打鐵趁熱，儘早把信回了。

　　顧客會洽詢，代表有購買意願，這種洽詢多半具有時效性，誰能保證他們的意願不會隨時間而消退？如果回信延誤或錯過回覆，商機也許會被對手搶走。

## 不要怕說對不起

　　有些公司怕說對不起，大概是因為他們認為一個企業不該輕易道歉。但公司本來就是由人所經營，犯了錯本來就該認錯，公司或個人沒有不同，何不人性化一點？在犯了錯或讓顧客失望時候，不加遲疑

地說對不起。

應該道歉的時候就該道歉，但不要過當，所謂過猶不及。

真誠而有效的道歉信需要具備四個條件：

1. 附和對方

2. 承認錯誤

3. 表示歉意

4. 提出補償辦法

左欄這封冗長的道歉信不太有誠意，激怒作用遠大於安撫作用。可以依據上述的四原則，把原信修改成像右欄這個樣子。

| 原信 | 修訂後 |
|---|---|
| First of all, Mrs. _____, we are truly and genuinely disturbed to learn from your recent letter that you are unhappy about the service you recently received from our Parts Department. The statement in our ___ _____ letter that the _____ were sent to you on _____ __ was entirely truthful. Without indulging in self-praise, we do make a sincere and honest attempt to ship orders promptly; but we have little or no control over shipments after they leave our place of business. Of that I am sure you understand, Mrs. ___ _____. Looking at this situation from your point of view, we can see why you have cause to be critical of our service.<br><br>Perhaps you would in the future prefer to send your orders to our _____ office. This office is set up, as I feel sure you can understand, to expedite shipments to our valued customers in the _____. However, | Dear Mrs. _____:<br><br>I don't blame you for being irritated about the handling of your recent order. It's exasperating to order something then not receive it.<br><br>It may have been our fault, but I hope not. Our records show the package was shipped on _____. It may have been lost in the mail that sometimes happens. Regardless of who was to blame, it's annoying and we're very sorry. The two order pads are enclosed. |

| 原信 | 修訂後 |
|---|---|
| as I previously explained to you, you were caught between our two offices, for which I am deeply sorry. But let's continue on a more positive note. Immediately upon receipt of your valued letter this morning, I personally went to the warehouse and pulled the enclosed _____. <br><br> Enclosed are two each of those pads which you requested. We value you both as a customer and as a friend, Mrs. _____, and we will do everything that is in our power to keep our relationship on a friendly basis. Won't you please write us and let us know, at your very earliest convenience, if everything is entirely satisfactory now, please? <br><br> With sincerest regrets, | We've enjoyed serving you for a long time, Mrs. _____ ___. <br><br> Sincerely, <br><br> P.S. For the fastest possible service, you might like to place future orders with our _____ office, which is nearer you. |

下面這封信除了道歉，還感謝對方的諒解，是個極佳的雙贏示範：

Dear _____:

Sorry that you had to wait such a long time for your credit of $_____. We have had trouble tracing your sale and the return of part of the merchandise.

The refund check has now been mailed. Again, thank you for excusing the delay.

Sincerely yours,

　　有過失時，勇敢承認並虛心致歉比任何巧言善辯都容易得到諒解，人與人之間是如此，公司與人、公司與公司間也莫不如此。現在來看看兩個值得推薦的例子，其一：

If it weren't illegal, Mr. _____,

I would enclose my head with this letter. We have taken an undue amount of time to reply to the above requisition. In defense of our tardiness, I might point out that the manufacture of the part involved is a more complex affair than one might expect.

　　其二：

Your letter of the seventh is on my desk, and to be perfectly frank, I goofed in entering your order and did not specify that it should be sent to camp.

　　上面兩段文字分別是兩封道歉信的第一段，寫信者直承失誤，真誠道歉，語氣謙和，收信者見信就已經釋懷。

## 善解顧客的抱怨

　　客戶難免會抱怨，有些抱怨真的很沒有道理，你若因此而生氣，要以牙還牙，那註定要把事情弄糟。對方以羞辱的方式來抱怨，你如果反唇相譏，只會把對方更激得火冒三丈，逞一時之快，後患無窮。所以，別讓一些抱怨影響你的情緒。

　　處理客戶抱怨的第一個原則，就是對那個抱怨不要太敏感，如果

你有被惹火的危險，不如把這件事交給別人處理。

對方抱怨，可能是因為需要同情，這時你該有寬大的心胸來安慰對方，告訴對方你若能力所及一定幫助，要是實在沒辦法，就委婉地讓對方知道實情，絕不要指責對方過分。一個人若不覺得自己過分，你再怎麼說都沒效，說他過分只有更糟，他反而可能更過分，這一來你豈不是更有得瞧。

收到一封無禮的來信時該怎麼做？

答案是：可以做任何有助於消氣的事，但千萬別在生氣的情況下回信。深呼吸把對方狠罵一頓，寫一封同樣無禮的信然後撕掉，直到你能夠沒有火氣地把那封信一笑置之之後，才正式回信。

處理這類信件時，記住三件事：

1. 寫信的這個人現在已經不像當初那麼氣了，他所有的怒氣都已藉著這封信發散出來，也許現在正對自己信中的魯莽感到羞愧。

2. 在給友善和藹的人寫信的時候，任何人都很容易保持良好的態度，而在寫信給無理取鬧者時還能保持良好的修養，就只有功力很高的人才做得到。

3. 不管對方生的是什麼氣，如果你體貼親切地回信，他要對一個友善的人保持生氣的狀態，幾乎是不可能的。

用信溝通歧見並不容易，需要熟練的人際溝通技巧。對無足輕重的小歧見別太認真，讓彼此更有空間。

採取大事化小的方式，世界上太多小題大作的人，實在該多培養大事化小的態度，即使對方想法偏差或情緒偏激，你也要冷靜以對，不要聞「激」起舞。一個巴掌拍不響，吵架不能解決的事情，冷靜的溝通往往能奏效。

要改變別人的看法有絕大的困難，若沒有必要，不要多事，這樣

可以避免更高層次的不和。

　　下面左欄這封信的態度就過於冷硬，不留餘地。寫信者雖有正當理由可不支付利息給存戶，但卻沒有說明那是基於銀行政策的不得已決定，結果容易造成不必要的爭執。不妨將信改寫像右欄那樣，雖然長了點，但卻比原信優秀得多，因為它有以下幾個優點：

1. 表現出對存戶的問題有相當的了解。
2. 說明了不支息的原因。
3. 說明了銀行的政策。
4. 消除了對方因為不了解銀行政策而提出的責備，使對方對銀行產生了友善的感覺。

| 原信 | 修訂後 |
|---|---|
| Dear Sir: | Dear Mr. ＿＿＿＿ : |
| In answer to your letter, we hereby advise you that in order for a savings account to earn interest, it must remain on deposit at this office for a full interest period, i.e., ＿＿＿ through ＿＿＿. | Your concern over why your deposit did not earn any interest between ＿＿＿ and ＿＿＿ is understandable. There is, however, a good reason. |
| If your funds had remained at this office through ＿＿＿, the interest would have been included, but since the funds | To pay interest, a bank has to invest its funds and earn interest. To do this, the bank must have the use of the money for a period of time. That's why banks generally observe definite interest periods-periods during which your savings must remain on deposit in order to earn interest. |
| | In our case the interest period is from ＿＿＿ to ＿＿＿. To earn interest, your funds would have to remain on deposit until the end of full interest period. |

| 原信 | 修訂後 |
|---|---|
| were withdrawn on ____ __, we could not pay you interest.<br><br>Very truly yours, | This is standard banking practice. I'm extremely sorry that it was never properly explained to you. If you have any further questions, please stop by my desk. I'd be delighted to help.<br><br>Sincerely yours, |

　　每個人的看法都有非關邏輯的個人成分在內，人如果不是自願改變，就連真理也不能使他改變，所以要想改變對方的看法，光靠邏輯或理論是不夠的。

　　任何人越被激怒，就越會固執己見，歧見應審慎處理，對在你之下的人，也應表現友善和柔和的態度，柔和得不會觸怒對方，從而也使對方的態度變柔和。

　　最好的策略在找出雙方的共同點，試著同意對方的看法，告訴對方，你也欣賞他們的一些觀點，儘可能把相容處找出來，最後才誠懇的說出你無法認同的地方，用誠懇的態度去做，對方就不會因為意見不合而不快。

　　你若有心，對方也會感受到。你的表現會改變對方的看法，你開誠布公的態度也能引發對方開誠相向，這樣一來你更能贏得對方的合同，乃至贏得對方的心。

　　不要在同一封信裡一下扮黑臉，一下扮白臉，信裡若有了黑臉的成分，你再怎麼做也沒法漂白。在需要謙恭有禮時，露出不滿或不悅的情緒只會帶來絕大的反效果。

避免敷衍的態度

有家商號在對街的銀行開戶，之後不久就收到該銀行發來的制式回信（Form Letter，或稱罐頭信、樣板信），信上除了謝謝他們來開戶之外，還建議他們若覺得銀行太遠，可以用郵寄的方式往來，由銀行負擔郵資。可是銀行就在對街，怎麼會太遠不方便？

許多處理信件的部門所在意的，只是儘快把來信處理完畢，而不是處理完善，所以公式化的信有時會鬧笑話，使用前要多斟酌。

有時候寫信給一些大公司的部門，得到的回答可能是：

> Sorry, that's not in our department. 或
> Not handled here ！

信上卻沒有建議接下來該怎麼做。

某客戶寄了一張訂單到某國際公司的紐約辦公室，卻得到這樣的回信：

> 　我們在紐約的辦事室沒有存貨，建議您寫信到總部辦公室詢問。

但信上卻沒有寫明總部辦公室的住址。

諸如此類，這種敷衍態度很容易毀損商譽及「錢途」，最好的預防方法就是不要輕易放過類似的錯誤。

處理顧客詢問的方式反映出你自己的態度，一種是敷衍推搪式，只想把事情推開，而另一種是盡心盡力幫顧客達到目的。

以下就是這兩種不同處理方式的樣本信，左欄只提供客戶進一步

的聯絡地址，請客戶自行聯絡，右欄則是主動幫客戶設法聯絡，顯然更易贏得客戶的心。

| 原信 | 修訂後 |
|---|---|
| Thank you very much, Mr. _____, for you letter of June 20 regarding _____.<br><br>Pleased as we are that you found this item quite interesting, we must regretfully inform you that our supply is just about exhausted and we have no plan to reorder just now.<br><br>However, if you wish, you may contact the following company which we are sure will be more than happy to supply you with the necessary quotation and information. It is: _____.<br><br>We're sorry that we cannot help you personally, Mr. Blank. Perhaps at some time future you will give us the opportunity to serve you .<br><br>Cordially, | Thank you very much, Mr. _____:<br><br>for your mail of 6/20 saying that you are interested in _____.<br><br>Unfortunately, we do not plan to continue carrying this item when our present stock is exhausted-in a few weeks at the most. To save you time, we are forwarding your inquiry directly to their distributor, _____.<br><br>You will hear from them promptly, I am sure. But please don't give up on us, Mr. _____. I hope the next time we can be of more service to you .<br><br>Cordially, |

　　商場上使用僵化的信件由來已久，部分原因是積習難改，部分的原因是寫信的人懶得用心，下面這封是某顧客寫給一家食品公司的信：

Since the product is delicious, I have tried to be understanding but last night was to much. We had guests for dinner and when one lady tried to shake some of your dressing on her salad, nothing happened. She then slapped the bottom of the bottle and a big gob shot out, burying her salad and, incidentally, ruining her dress. She was upset and I was embarrassed, so I thought I would write and suggest that you start putting this dressing in the same wide mouth jar you use for mayonnaise.

不久後該公司用呆板得可以的方式回信如下：

In reply to your February 16 letter we hereby advise that your suggestion has been forwarded to our package design center for their consideration. We appreciate your interest.

別人在書信往來時若都這樣沒有新意，只要你與眾不同地寫出些充滿人味的信，一定能夠在人際關係和公務流通上造成一股清流，千萬別寫得像下面這封信一樣，充滿廢話：

With reference to your letter of ＿＿＿＿ requesting the names of everyone from our agency who are planning to attend the regional conference and also requesting as to how the names should appear on the name badges.

Enclosed herewith you will find list with reference thereto.

有些回覆客戶洽詢的信沒能抓住重點，以致答非所問。要避免這個毛病，最好的方法是養成仔細審閱的習慣，在回應前先找到問題重心，簽名寄出以前再檢查一次，確定充分照顧到了每一點。

來信者若問了很多問題，記得要逐一回答。同時，除非和對方問題有關，不用另生枝節，提供超出範圍的資訊，造成徵詢者不必要的困惑。例如：客戶洽詢某商品的報價時，不該提供不相關的價目表，讓對方花時間找出他們真正要的資料，應該要為對方標出重點。予人方便，自己方便，雖然要多花點時間，但是你的服務必然會受到肯定。

若只把答覆洽詢當成例行公事，這樣的態度對業務沒有好處。來詢問商訊者多半有考慮要消費，他們通常期盼能得到好的回應，因此，顧客詢問的信件應妥善處理。

##  注意回信的情緒

處理的投訴書信多了，有時會碰上火氣十足，帶有侮辱性的來函，這時切忌反唇相譏，不妨視若無睹，如果情勢所逼需要回覆，也該先冷靜下來。通常對這類的信報以禮貌的態度，針對事實回覆是最好政策，這樣可以免得引起進一步的辱罵或筆戰，省時省力。有時對方還會因你以禮相待而自覺汗顏，自動就自己的不當言論提出道歉，這是經驗之談，宜加三思。

因此，心情不佳時暫勿寫信，如果執意要寫，你的心情一定會反映在信上，即使你自己察覺不出，讀者也不難由文章中發現你的不滿。

下面是一位自認受委屈的業務員所寫的信，糟的是寫信者沒有察覺自己這封信具有攻擊性，他喃喃抒發自己的不幸，而沒有理會到客戶的感受。

Dear Mr. _____:

I don't know if you realized it or not, but I made a special trip out of my way to go down to Tainan that day to see you and to give you first hand an explanation of the service that we have to offer. Your letter previously sent to our Taipei office made this necessary because you gave them the idea that I was not adequately covering this territory.

I would therefore appreciate it if you could bring yourself around to make a decision in this matter which otherwise would coast along for lack of initiative. If you have not already spoken to our good subscriber, Mr. Chou of your city, I suggest you do so promptly, and thus arrive at a beneficial conclusion for yourself.

Enclosed kindly find order blank. Kindly fill out and return immediately.

Yours sincerely,

　　在處理公務信件時，你偶爾會想寫信狠狠教訓對方的不上道，如果你有這種衝動，千萬先忍住，對方再怎麼無禮，你也沒有必要還擊，而大可用感性或幽默的方式來回應，所謂和氣致祥、和氣生財。

　　雖然我們的原則是儘速回覆，但是在心情欠佳的時候，還是先靜下來，讓時間調和一下心境之後再行動筆，能夠這樣，誤事的可能性一定大大減低。

　　下面這封信就很有責怪意味，不值得取法：

Gentlemen:

We note that you deducted an unearned cash is count of $_____ when making your recent payment. _____ Corp. offers a cash discount for payment of an invoice within _____ days of the invoice date. This offer isn't extended because we want to give money away. It isn't an additional discount off the price of the product. Future terms must be more restrictive if unfair advantage is taken respecting cask discounts. I'm sure you understand.

處理信件的時候，別只顧效率而忽略友好和親切。在一下子要回覆很多信的時候，若開始迅速處理，友善和親切感就容易消失，爲了挽救這種缺憾，有時必需把信重寫。

那麼該怎麼寫呢？

友善的信可以造成極佳的印象，給收信者好印象是絕對有用的，即使你要上法庭告對方，也可用友善的態度來表達，告訴對方你之所以訴諸公堂也是不得已的。

千萬不要像下面這兩封信一樣，廢話連篇。

其一：

We wish to advise that we will endeavor to determine objectively the overall consequence of failure to adhere rigidly to established precepts in connection with activities in the branch office to which you are currently attached and it is emphatically stressed that you direct your primary effort to the end that basic circumstances be altered with respect to current concepts of preferential routines.

其二：

In accordance with criteria contained in paragraph 4 of above reference and in consideration of duties to be performed by course graduates the following member of this command is nominated for attendance during the period 8-Jan ~ 22-Feb.

這些信都是實際的案例，絕非創造出來作負面示範的樣本，它們全都出自受過相當教育的人之手。事實上，沒受過足夠教育的人也寫不出這些疊床架屋的「巨作」來，這些信可以說是糟得沒法修改，只能全盤重寫，大家不妨試試看，用我們一貫強調的方式來重寫這些信。

## 婉拒的藝術

你也許會有不得不拒絕顧客要求的時候，在這種情形下，你若還想和他們保有良好的關係，就得用友善的態度來製造有利的形勢。

請看左欄這封婉拒客戶要求退貨的信，寫信者已盡量做到婉轉的原則，但信中否定的語氣依然太過明顯。顧客雖非永遠是對的，但他們有所要求時，你還是應盡量配合，如果實在不合理，你也不宜過於失禮，大可表現出你很想要答應他，但不是你不答應，而是實在做不到。準此，原信可以改寫如右欄。

| 原信 | 修訂後 |
|---|---|
| Dear Mr. _____ : | Dear Mr. _____ : |
| We're delighted to hear that your copies of "How to Drown-proof Your Family" have been located. | We're very sorry to learn that the copies of "How to Drown-proof Your Family" were mislaid in your warehouse. I wish we could take them back or help you dispose of them elsewhere. Unfortunately, they are imprinted with your company name. |
| Unfortunately, we cannot allow a partial credit on these booklets. Imprinted booklets are of no value to us on a return. | Has it occurred to you that the booklets could be used for an identical promotion next year? The content is timeless it will be every bit as good then as it is now. |
| Your honoring of our invoice No. 7141237 will be greatly appreciated. A copy of the invoice is enclosed. | I'm sorry we can't be more helpful. |
| We regret our inability to be more helpful to you. | Sincerely, |
| Sincerely, | |

## 一、婉轉的重要

　　寫拒絕的信不太容易，草率的寫法不易讓讀者心服，惡劣的態度絕對令人詬病，因此你想要拒絕別人的請求時，千萬要婉轉一點。

　　左欄這封拒絕求職者的信，對於求職的人相當冷酷，沒有給求職者留餘地，這樣的方式非常不妥。任何人很有敬意地表明工作的意願，公司即使不想雇用，也應以禮貌的方式回覆，這樣不僅保留求職者的自尊心，公司也會有較好的形象。右欄封信就是很好的範例，該信婉拒了求職者，但隨即稱許了他的能力和資格，而且還祝福他，乃是達成雙贏的好方法。

| 原信 | 修訂後 |
|---|---|
| Dear Mr. _____ :<br><br>Regarding your application for employment with this company, we have nothing to offer you.<br><br>Very truly yours, | Dear Mr. _____ :<br><br>Thank you for your application of ____ __. I am very sorry that we do not have any vacancies now, and do not expect any soon.<br><br>With your excellent qualifications, you should have no difficulty finding a suitable opening. Wish the best of luck, Mr. _____.<br><br>Sincerely yours, |

下面這封信件也充分使用了婉拒的藝術：

Dear Mr. _____ :

Thank you very much for your application for employment.

All the positions in our organization requiring the services of a man with your qualifications are, unfortunately, filled.

We are glad to know of your interest in becoming associated with _____, and would like to retain your file letter and data sheet in our filed in case something unexpected should develop.

Sincerely yours,

　　要拒絕一項請託，記得態度要柔和，讓對方了解是事實不允許，並且向對方道歉，說明不是你不願意，而是目前無能為力。千萬別明指或者暗示對方要求過分，因為沒有人會願意受到折辱。

## 二、客戶的信用

若有信用不好的客戶要和你做買賣，你可以用委婉的方式，以如下的信件告訴對方：

Thank you, Mr. _____,

for your order for _____ which you placed with Mr._____ on _____. Unfortunately, we do not have in our credit file information which would enable us to make this shipment on open account. As soon as we do, it will be a pleasure to extend you our customary trade terms. In the meantime, if you will send us your check for $_____, we can ship your current order immediately.

這封信值得效法的地方，在於沒有直說顧客的信用不好，而只說自己沒有足夠資料來了解對方的信用狀況，只好請對方先行付款再行交易，沒有觸及客戶的痛處，簡短扼要，沒有因多做解釋而有越描越黑之虞。

# 第七章
# 催收帳款的心理學

本章提綱

　　◎威脅敵意不是辦法
　　◎溫和體諒效果不差


　　寫催款信（collection letter）時，我們除了考慮是否能拿到錢，還希望能不打官司，也不令客戶起反感。要想到，今天拖欠款的客戶日後生意壯大了，可能變成你很好的客戶，犯不著現在得罪人家，以致自殘商機。

 ## 威脅敵意不是辦法

　　催收不是件愉快的事，傳統的催款信有時不是太友善，請看下例簡直是充滿了怨毒跟敵意：


Gentlemen:

This is to inform you that we have carried your account just as long as we intend to-far longer than you deserved. Unless it is paid in full immediately, we intend to turn this over to our lawyer for prompt action. I just can't understand people like you ignoring statement after statement and letter after letter. Surely you have some sense of business honesty. If so, kindly send us that

$_____ you have owed us for the past ten months or you'll hear from our lawyer.

Yours very truly,

催收的目的在於說服對方付清欠款，因此其中心主旨應是：

1. 請欠方了解自己以誠相待的立場。

2. 請欠方了解付清欠款可以省下許多麻煩。

3. 請欠方付清欠款。

千萬不要用下面這種方式寫信：

Apparently you choose to ignore all mail regarding your bill and your obligation to pay it.

In good faith we sold merchandise and services to you and you signed an agreement when you asked for credit, because you were without funds to pay cash I presume, and now you refuse to pay for the merchandise.

This is an excellent way to obtain the material things in life we wish to impress our friends with, but hardly honest. Most of us have to pay for the things we obtain in life, though it would be much simpler you way.

Let's get this balance cleared and the money in this office now! I believe we have been patient quite long enough and the time for excuses is past. Please remit the full balance promptly.

Very truly yours,

　　左欄那封收帳信的最後一句是敗筆，但在那類信裡竟然常出現這種句子，言外之意像怕不付錢，這種語氣很可能造成對方不爽，何不用像右欄那樣以較讓人接受的方式表達？

| 原信 | 修訂後 |
| --- | --- |
| Thank you for your order for six gross of pencils @$_____ a gross. The pencils are being shipped today. | We look forward to receiving your check for the proper amount in the next few days. |
| We are returning, however, your check for $_____ which accompanied the order. The correct extension of this price is $_____. | The check for $_____ which accompani-ed your order is greatly appreciated. Checking the extension, however, we find it should be for $_____. The enclosed invoice credits you with the amount already paid, leaving a balance due of $_____. |
| We look forward to receiving your check for the proper amount in the next few days. | It's a pleasure to have the opportunity to serve you. |

## 溫和體諒效果不差

　　溫和的訴求比威脅的態度，更容易讓人償還積欠，因此下面這封信的機會就比較好：

Gentlemen:

I am sorry you have not answered my previous request for payment if your long overdue account of $_____. I have tried in every possible way to get your side of the story.

You received our merchandise, and we naturally want our money. Although our lawyer, Mr. Smith, is a very good collector, we don't want to get our money that way unless we have to.

So I'm making one more appeal to you. Don't you think it only fair that you send us this money? We have waited a long time, you know. Think it over, and then drop a check in the mail today-PLEASE!

Hopefully yours,

下面這封友善的討債信第一次寄給 626 個欠債戶就收回了 262 項（44%）欠款：

Dear Mr. _____ :

Yesterday our treasurer called me into his office and said: "I see that Mr. _____ has not yet settled his account. In fact, he hasn't made a payment on it since for _____ months, though I've written him several times. I did not wish to bring suit, for they've had pretty hard times in that section during the past year.

"Now, however, conditions are better there. I'd like you to write to Mr. _____ and ask him to clear up this account. We've been fair with him, and I think you will find that he will want to be equally fair with us."

I thought I could do no better than to tell you just what our treasurer said to me. We have waited a long time, you know. So I am going to ask you to write and let me know just what you can do for us.

Yours very truly,

餘下的 364 個尚未付款的戶頭在收到下面這封信之後，又有 165
戶（44%）還清：

If a customer owed you $_____ and for two years had paid nothing on it, how would you feel?

But now suppose you knew that that customer had been up against hard conditions all that time. You put yourself in his place and decided not to appeal to the law to collect your money. Then when things picked up with the customer, suppose you wrote to him as man to man, asking him to treat you as fairly as you had treated him. Wouldn't you feel certain that, as a businessman and as a gentleman, he would respond? Wouldn't you?

There are laws that regulate business, Mr. _____. But the biggest thing that keeps business clean and aboveboard is the fact that most men believe in fair play. Business would go to smash if we couldn't depend upon the sacredness of a commercial agreement.

That's all we ask from you, Mr. _____, a square deal. You believe in that, just as we do, don't you? Then let's settle this debt as between friends and gentlemen.

A check from you would confirm our belief that you do believe in the square deal.

第三封信還是沒有擺出討債相，寄出之後再收到 165 戶償還欠款，證明了用蜂蜜比用毒藥更容易抓到蒼蠅。

　　以上的欠款都是積了 2 到 5 年，用其他方法催了又催也不見反應的，可見這類的信確實很有魔力。

　　有效的催款信多摒除威脅的方式，想不對簿公堂就讓對方情願還錢時，該怎麼做呢？下面有幾個原則可供參考：

1. 要友善，不要惡毒 —— 抨擊只會使人厭惡，只會減低收款的機會。不管你最後會採取多激烈、多不愉快的行動，都該以哀矜勿喜的態度出發，就像對待朋友一樣。

2. 要體諒對方，不要板起面孔 —— 體諒會增加付款的意願。因為許多人就算有錯，也不想挨訓或受罰，所以不要教訓對方不付款。

3. 假定對方只是失察，不是有意 —— 很多情況下人家確是失察，若你把失察當成故犯，難免激怒對方。

4. 說些好話稱揚對方的好處 —— 溫言細語比別種方式更能激發人家還錢的意願。

5. 不要威脅 —— 如果被迫採取法律途徑，也最好用不得已的態度通知對方，不要用強烈的方式來威嚇。

　　這裡有一封有效的催帳函，沒有一絲責罵的語氣：

When we do not receive payment within a month or so for books shipped to a customer, we wonder...

Did he get what he ordered?

Did he forget that the free examination period was for only ten days?

Did he overlook the invoice enclosed in the shipment?

This is just a reminder that we have not as yet received payment or heard from you since we mailed your order. May we hear from you promptly, please?

　　下面這個範例也非常成功：

If you and I could meet in person and shake hands, I'm sure that the first thing that would come to your mind would be the $_____ you owe me.

Then don't you think it only fair to ask you either to pay this bill now, or to sit down and tell me why you haven't taken care of it?

I certainly want to be fair with you. I think I have been. And I'm sure that you want to be equally fair with me. Am I right, Mr. _____?

　　幽默感可以不用造成傷害，是取得對方合作的好方法，下面這封信就是個很成功的例子：

We want a check of some kind!

Either a real check, or a pencil check alongside one of the items listed below. We would like to know exactly where we stand, so just check up on your bankbook today and then drop a real check into the mail tomorrow. Or check one of the blocks below and drop this letter into the nearest mailbox tonight, using the en-closed stamped envelope.

I am sending check herewith

Here is part of your bill to show that my heart's in the right place.

I'll try to pay each month from now on in the same amount as the enclosed check.

I think I can pay this on the _____, so here is my postdated check.

Here's all of it-now SHUT UP!

　　最後這封信旨在對債務人發出最後通知，但它既不惡毒也沒有抱怨：

The time has come when we must write the letter we dislike very much, Mr. _____.

In our previous letters we said about everything we could think of in urging you to pay your long overdue bill, which is now $__ ____, including late charges and interest.

But you have made no replies to any of our requests.

The only thing left for us to do is to turn your account over to our attorney for collection. Although he is very successful in getting payment, we don't want to do this without giving you one last chance to write and let us know your side of the story. For that reason we will delay action for five days—until April 10—before turning your account over to a collection agency.

We have been good business friends for many years, Mr. Doe. Please don't make it necessary for us to terminate our friendship in such an unpleasant way. Write us immediately—or better still, send us your check for at least a part of your bill. But do it without delay, please!

　事實證明，寫信要帳時，使用禮貌的態度比較容易成功。下面這封也是一封不錯的催討信：

Good morning, Mr. _____:

My boss has run out of patience, and I've run out of excuses. Thus, when he called today, my back was against the wall. Certainly, these are my troubles, and not yours, Mr. _____, but in this particular case, you're the one man who can help me. I've gone out on a limb and promised a settlement of your account by the end of this month.

I know this has been a problem for you, but I think you now have the necessary information to reach a settlement. In the event your files are not complete, I'm enclosing a list of all missing and damaged equipment.

If you'll just let me know what time would be most suitable for a meeting next week, I'm sure we can work out this settlement in just a few minutes. Then we can both forget about it and my boss can stop feeding his ulcer again.

大家可以注意到上面幾封信裡有幾個要點：

1. 盡量體諒對方，往好處想。

2. 絕不用教訓的口吻。

3. 若萬不得已要對簿公堂，也不妨以友善的態度通知對方。

以下有幾封有效的催款信，沒有一封用譴責的方式訴求，債務人看完後多有很好的回應。

其一：

Your account, Gentlemen—

shows a past due item of \$_____ which is represented by our in-voice #_____ dated _____.

Is it because of some omission on our part that we haven't received your check? If so, please write us the details and we will take any necessary action at once.

If on the other hand, our account is ready for settlement, a check at this time will be appreciated.

Thank you.

其二：

Suppose, Gentlemen,

we owed you a sum of money and it was long overdue, what would you do?

We believe you would ask for it. If payment were not made within a reasonable time, you would no doubt ask us why.

We don't know what has caused the delay in payment of your invoices for $_____, but since it is now _____ days past due, don't you think it should be paid without further delay?

Please let us hear from you.

其三：

Your attention, Gentlemen—

—to your past due account will be appreciated. The balance of $_____, represented by _____ invoices is now _____ days past due.

We have not been demanding in our requests for payment: you can return this courtesy with a sincere effort to settle your account.

Thank you.

其四：

Just a reminder

that your account is past due. Perhaps you have already mailed a check for the $_____ invoice. If so, please consider this a—thank you.

其五：

Statements and letters, Gentlemen,

have been sent to you regarding your long-overdue account for $_____.

In reviewing the file, it appears that every opportunity has been given you in which to arrange for settlement.

However, before placing your account with a collection agency, we request once again, your check before _____, please.

　　向人催帳而不令人感到不悅是一種藝術，請再看下面兩個有效的例子，其一：

To be frank, Mr. _____, I was racking my brain for some clever new way to put this idea across when it occurred to me that you would respond to a direct request for payment. So here it is:

Your loan is past due for the sixth and seventh payments. So won't you please send us your check today? Thanks very much, Mr. _____.

　　其二：

When an account runs past the due date, we find that most of our good customers appreciate a brief reminder. Consequently, this note is sent to call your attention to the overdue amount shown on your last statement.

If you haven't sent us your check already, won't you please mail it today for sure?

　　有時你可以提醒債務人，良好的債信是很珍貴的資產，債信一旦破產就很難恢復了。

We considered it a real privilege and pleasure to open an account for you on _____. That's why we're puzzled by your not sending us your regular monthly payment when it was due on _____.

None of us likes to get off on the wrong foot, particularly where our credit is concerned. A good credit reputation is one of the most valuable possessions a person can have. Once lost, it can be extremely hard to re-establish.

There may have been a number of reasons why this payment was not sent. It was your first and probably you just overlooked mailing it. Won't you please reaffirm our confidence in you by mailing your check today—then by making future payments as agreed?

　　對方不付款的原因很多，可能只是一時疏忽，也可能手頭緊，或不滿意你那批貨，或在某方面對你不爽，甚或是真的要賴帳。

　　不管什麼原因，用友善和體諒的態度站在對方的立場規勸，提醒對方其債信或商譽可能受損，進而指出你在萬不得已時可能提起訴訟。

第八章

# 報告與通告

本章提綱

　　◎本文合邏輯

　　◎複述要有原則

　　◎通告或通知性的文件

　　◎注意文章組織


　　報告或通告的主要目的，是要傳達資訊或回答問題，你若有撰寫報告的機會時，不妨把報告當成一種特別的信，依照前述的寫信原則來寫的報告或通告，可讀性和實用性都會增加。

　　若你需要把某調查結果寫成報告，同時為上司擬出行動方案，應該從那裡著手？這時不妨設身處地，想想看上司會希望從哪方面開始？

　　上司應該不想在讀了好久之後還找不到主旨，因此報告必須簡單明瞭，避免浪費讀者的時間。如果報告的頁數多，你只要給他最重要的就好了，在開頭處列出重點或主管摘要（Executive Summary）。

　　寫給任何人看的報告都一樣，開頭的摘要非常重要。

## 本文合邏輯

　　線性的邏輯指一貫而不旁生枝節的邏輯方式，乃是良好報告的基石。報告的主體應該涵蓋事實、結論和推理的方式，完整而邏輯地從引用的事實導出最後的結論。

要把這結構架設出來，那就得先花時間理出報告的要點，一個接一個，從開始到結束，環環相扣，由前一要點引入後一要點，上一要點引入下一要點，遵循線性的邏輯直到導出結論為止。

注意下面幾個要點：

1. 避免不實和吹噓。

2. 把事實和意見分開，盡量不要加入個人的意見。

3. 可以多使用輔助圖表，讓文章看起來生動活潑。

4. 版面與編輯的方式宜多留些空白，這樣會讓文章容易閱讀。

5 避免繁瑣的文字，要用簡短的文句和清晰的思路，帶給讀者深刻的印象。

##  複述要有原則

要因應讀者的立場，考慮讀者的吸收能力，揣摩讀者的心意，讓讀者能不費力就明白你的意思。除了文章要完整清楚，還要避免為了增加篇幅而硬塞些不必要的資料。

只須表達必要的訊息，不太重要的題材不用全盤托出，最好讓讀者有時間把先前的主題都消化得差不多了，才搬出下一個主題。

「重述」可以讓讀者有複習的機會，有技巧的作者會用稍為更動過的寫法，把他們想讓讀者吸收的主題加以重述。

下面是幾個適合使用複述原則的地方，在這些地方可以用不同的複述技巧，像：舉例說明、互相比較，或用修改後的句子重新描述一遍等。

1. 不夠清楚的地方。

2. 需要加強語氣或強調重點的地方。

3. 一連串的主題需要加深讀者印象的地方。

　　只要覺得任何地方表達得不夠清晰，就不妨想個能讓讀者掌握主題的方式，來加以舉例或比較。

　　以下的兩個例子是重點複述的代表作，加底線的部分，其目的是要進一步突顯文意，並加深讀者對前文的印像。這種做法的好處在短文裡還不太明顯，但在長文裡就顯出來了。讀者若在前文中看漏了東西，還有機會在後文的複述中複習一遍，不用回頭費力搜尋。其一：

　　Continued learning is important not only to your health and happiness but to your economic security. Times changes, so do job requirements. Your best guarantee of employment is your ability to change with them. <u>If you stay flexible—ready, willing and able to learn new things—you have nothing to fear from age or automation.</u>

　　其二：

　　As your head comes above the surface, a slightly stronger push with your hands will hold it there long enough to inhale slowly through your mouth. <u>Take your time inhaling</u>, don't gulp.

　　下面是一篇還不錯的考察報告，其中列出了考察項目的性能與運作狀況，簡單扼要，不帶任何無關的資料，讓讀者不用多花時間就能掌握重點。

Equipment Offered For Sale At ＿＿＿＿

＿＿＿＿ and I visited ＿＿＿ to inspect the mining and milling equipment.

The merchandise that appeared to be in first-class condition were ＿＿＿ mills. The mills are driven through herringbone gears by direct-connected, slow-speed electric motors. The ball-mills are 6'x 6', 5'x 5', and 5'x 6' in size. The 6'x 6' mill appeared to be in the best condition, and was located in a position where removal from the plant would not be extremely difficult.

＿＿＿, liquidator, has reduced the price on the 6'x 6' mill from $＿＿＿ to $＿＿＿. We offered $＿＿＿ for the 6'x 6' mill complete with motor, starters, apron feeder, and miscellaneous spare parts loaded on our truck at the mill site.

Mr. ＿＿＿, the representative of ＿＿＿, will call you on ＿＿＿ with their decision relative to acceptance of our offer.

## 通告或通知性的文件

在有些人心目中，發布通知是件正式的事，因此他們認為通知信、通告函或布達性的文件就該嚴肅一點，有這種想法的人寫起通告來容易僵化。

左欄的例子，作者用正統的公務書信文句來寫，雖然還不算太差，但讀來沒有親切感，因此算不上好。要讓通告與眾不同、吸引讀者眼光，最好是用閒適的心境來溝通，寫的人要是能放鬆心情，讀者心情也自然輕鬆，如同右欄的通告函效果會更好。

| 原信 | 修訂後 |
|---|---|
| Dear Buyer: | Good News, Mr. Buyer! |
| We would like to take this opportuni-ty to introduce our organization to you. Last week Mr._____ very kindly wrote you advising that we are now the exclusive manufacturers and distributors of _____ ___. | _____ is your new supplier of _____. Have no fears—there will be no interruption in your service or deliveries. As exclusive manufacturer and distributor, we'll continue to supply them at the same terms and prices as previously. The service on your orders will be as prompt as ever. The only change is that all future shipments will be FOB our factory in ___ ___. |
| _____ is a division of _____, nationally known manufacturer for over _____ years. Our manufacturing resources and styling know-how, will bring to you the newest fabrics and most wanted colors. All this at the same terms and prices as previously sold to you. | _____ is a nationally known manufacturer for over _____ years. We plan to use our styling and manufacturing know-how to bring you the newest materials and most wanted colors. |
| Our new line is now being styled. We will send you this illustrated information within the next month. | Watch for illustrated information on our new line. It will be sent to you within the next month. Meanwhile, if you possibly can, stop in at our showrooms. I'll be there, looking forward to meeting you. |
| All sales will be supervised personally by Mr._____ at our showrooms. | |
| Until we have the pleasure of personally meeting with you, please accept our thanks for your continued patronage. You can be sure we will favor you and your organization with the very best service. | Here's to a fine relationship, one that will bring you even greater sales and profits on these two fast-moving items. If we can help you in any way or if you have any suggestions, please let me know. |
| Very truly yours, | Sincerely, |

把讀者當成朋友，是可放諸四海皆準的通則，這樣寫來的書信，多會誠懇而有效。把你的書信都用上述的通則加以檢視，如果還沒有達到標準，不妨重寫一遍，你會發現寫通知函也並不困難。

## 注意文章組織

寫作重要的書信或報告時，不妨把大略的構想分段打在電腦上，再把這些段落分類和編組，想想看：

1. 哪個構想該排在第一？

2. 哪個構想該跟隨在後？

3. 哪個構想該和哪個放在一起？

這樣自然對寫作流程有幫助。

然後問問自己，從哪方面著手會使讀者感到有興趣？試從有趣的角度來切入，用個正向和愉快的開頭提醒讀者：好戲在後頭！

文章如果太長，就要不斷燃起讀者的閱讀慾，想想看：

1. 哪些事情很重要？

2. 哪些事情容易吸引讀者？

想到了就加到文章裡。

這個方法只要不破壞文路，好好使用就可適時引發讀者的興趣。

「分段」是迎合閱讀心理的技術，分段後的文章在外觀上讓人較願意讀下去。左欄這篇文章未經分段，顯然讀來不易。所以，不妨善用分段的藝術，也就是：

1. 在文意轉折時應該分段。

2. 要吸引讀者的注意時可以分段。

把該信分段重整，成為像右欄那樣更為易懂。

| 原信 | 修訂後 |
|---|---|
| Dear Mr. _____ : | Dear Mr. _____ : |
| I am glad to send you, as you asked in your letter of Nov. 10, my version of the accident which occurred at ____ __ and _____ last Thursday evening at 6:4 Shortly after the traffic light turned green for north- and south-bound cars—I presume 10 or 11 cars had crossed the intersection—an old woman started to cross the street from the northwest corner. As nearly as I could see, she walked into the side of a southbound car a little ahead of me and just to my right, and was knocked down. I stopped my car and went to her. When I reached her, she was unconscious, but regained consciousness by the time I lifted her into my car. With Officer _____, No. _____, who was handing traffic at the corner, I drove her to the emergency ward at the _____ Hospital. When she got there, she was conscious, though somewhat dazed, and was bleeding slightly from a cut on her head. Shortly after we reached the hospital, the man whose car was directly involved the accident arrived and gave his name | I am glad to send you, as you asked in your letter of _____, my version of the accident which occurred at _____ and _____ last Thursday at 6:45 pm.<br><br>Shortly after the traffic light turned green for north- and south-bound cars—I presume 10 or 11 cars had crossed the intersection—an old woman started to cross _____ from the northwest corner. As nearly as I could see, she walked into the side of a southbound car a little ahead of me and just to my right, and was knocked down.<br><br>I stopped my car and went to her. When I reached her, she was unconscious, but regained consciousness by the time I lifted her into my car.<br><br>With Officer _____, No. _____, who was handing traffic at the corner, I drove her to the emergency ward at the ____ __ Hospital. When she got there, she was conscious, though somewhat dazed, and was bleeding slightly from a cut on her head. Shortly after we reached the hospital, the man whose car was directly involved the accident arrived and gave his name and address to Officer _____. |

| 原信 | 修訂後 |
| --- | --- |
| and address to Officer _____ . I hope that these details will give you the information you want. At all events, it is as complete a version of the accident as I can give. | I hope that these details will give you the information you want. At all events, it is as complete a version of the accident as I can give. |

　　分段後，這篇文字讀起來就少了「文字過度緊密」的壓迫感，讀來真的容易許多。

　　過長的句子與不分段的文章帶給讀者的感覺如出一轍，當然也有改進的必要[1]。

---

[1]　參看第 4.2 節「避免長而無當」。

Volume

# 3

# 謀職寫作體例

本篇提要

# 第九章

# 謀職與履歷

本章提綱

◎職涯發展的先期作業

　分析專業需求

　建立核心能力

◎英文履歷必備──Action Verbs

◎履歷要件解析

　基本資料（Personal Data）

　專長（Qualification Summary）

　經歷（Work Experience）

　　工作經驗描述要領

　　經歷豐富者的寫法

　　職場新鮮人的寫法

　學歷（Education）

　其他項目

　　應徵職務（Position Desired）

　　介紹人（References）

　　期望待遇（Preferred Salary）

◎完整履歷的範例

 職涯發展的先期作業

　　你可以依熱門行業的起落來調整自己進修的方向，但有些職業或職位不是市場上既存的，開拓性的思考能為你在職場上創造新的工作，或規劃出更有意義的職涯。

　　個人的生涯管理除了依照市場導向，也要考慮性向上的個別差異，所以你不妨盡量發揮潛能，把個人性向與事業融合。

一、分析專業需求

　　依卡內基訓練的定義，個人核心能力乃是指溝通能力（Communication）、領導能力（Leadership）和團隊合作能力（Team Work），這些核心能力是任何行業都注重的能力，是許多企業衡量用人的重要關鍵。

　　因此，具備個人核心能力的人，是任何行業都會需要的人才。

　　根據 CHEERS 雜誌 2004 年「臺灣 1,000 大企業最愛大學生」調查，企業挑選新人的考慮標準依序是：

　　1. 學習意願強與可塑性。

　　2. 穩定度與抗壓性。

　　3. 專業知識與技術。

　　4. 團隊合作。

　　5. 國際觀與外語能力。

　　6. 解決問題的能力。

　　7. 創新能力。

　　其中的第二、四、五等項，就分別是領導能力、團隊合作能力和溝通能力的指標。

二、建立核心能力

Daniel Goleman 在他的名著 *Emotional Intelligence* 中表示：

個人成功的關鍵，情緒智商（即 EQ）占 80%，而專業能力只占 20%[1]。

一般企業的「員工十大核心職能」中，與 EQ 相關的項目包括位列：
1. 第一的團隊合作（占 91%）。
2. 第二的主動積極（占 79%）。
3. 第四的責任感（占 70%）。
4. 第六的正直誠信（占 50%）。
5. 第七的客戶導向（占 50%）。

任何準備進入職場，或想在職場上爭取更多機會的人，除了要多花時間培養上述那些與 EQ 有關的核心能力外，一般人一生中最重要的投資，可能是花在建立良好的人脈關係上。

建立人脈主要應從平常做起。在學習與培訓、人際網路及各式交流的途徑中主動與人交談或向人討教，都是建立人脈的好機會。

## 英文履歷必備 —— Action Verbs

寫作英文履歷時，要盡量使用具有行動力的動詞（Action Verbs），使用這種動詞開頭的句子顯得比較正面而有力。

下面各節範例中用到的「Commend, Conduct, Develop, Manage, Maintain, Perform, Provide」等，就全是具有行動力量的動詞，由於履

---

[1]   Daniel Goleman, *"Emotional Intelligence"*, Bantam Books, New York, 1995

歷中描述的大都是些過去的經驗，這些動詞在例子中都以過去式出現：

Commended, Conducted, Developed, Maintained, Managed, Performed, Provided

而如果是談到當前還在進行中的工作，就要用進行式來描述：

Commending, Conducting, Developing, Maintaining, Managing, Performing, Providing

表 9-1 到 9-9 中所列的，就是適用於下列不同領域的 Action Verbs：

1. 溝通能力類（Communication Skills）
2. 行政能力類（Clerical and Detailed Skills）
3. 創作能力類（Creative Skills）
4. 財務能力類（Financial Skills）
5. 輔導能力類（Helping Skills）
6. 管理能力類（Management Skills）
7. 研究能力類（Research Skills）
8. 教學能力類（Teaching Skills）
9. 技術能力類（Technical Skills）

各表所列皆為各動詞之現在式，使用時宜依需要改為過去式或進行式。

表 9-1 溝通能力類的 Action Verbs

| address | 演講 |
|---------|------|
| arbitrate | 仲裁 |
| arrange | 安排；籌畫 |
| author | 著作；創作 |
| correspond | 對應；符合 |
| develop | 發展 |
| direct | 指導 |
| draft | 草擬 |
| edit | 編輯 |
| enlist | 徵召；應徵 |
| formulate | 陳述 |
| influence | 影響 |
| interpret | 解釋；翻譯 |
| lecture | 演講；授課 |
| mediate | 斡旋；調解 |
| moderate | 協調；主持 |
| motivate | 刺激；給予動機 |
| negotiate | 談判；協商 |
| persuade | 說服 |
| promote | 促進 |
| publicize | 公布；發表 |
| reconcile | 調停；使和解 |
| recruit | 徵募；補給 |
| speak | 演說 |
| translate | 翻譯 |
| write | 書寫；寫作 |

表 9-2　行政能力類的 **Action Verbs**

| approve | 批准；核准 |
|---|---|
| arrange | 安排；籌畫 |
| catalogue | 編目 |
| classify | 分類 |
| collect | 收集；集成 |
| compile | 編輯；編譯 |
| dispatch | 派遣 |
| execute | 執行；實行 |
| generate | 引起；產生 |
| implement | 實施；履行 |
| inspect | 檢查；檢閱 |
| monitor | 監測；監視 |
| operate | 操作；經營 |
| organize | 組織 |
| prepare | 準備 |
| process | 處理 |
| purchase | 購買 |
| record | 記錄 |
| retrieve | 檢索；恢復 |
| screen | 篩選；審查 |
| specify | 指定；詳載 |
| systematize | 系統化 |
| tabulate | 表列 |
| validate | 確認 |

表 9-3　創作能力類的 Action Verbs

| act | 行動；扮演 |
| --- | --- |
| conceptualize | 概念化 |
| create | 創造；產生；建造 |
| design | 設計 |
| develop | 發展；發達；進步 |
| direct | 導演；導播；指示；指揮 |
| establish | 建立；制定；確立 |
| fashion | 形成；塑造 |
| found | 發現；建立 |
| illustrate | 舉例 |
| institute | 創立；開始；制定 |
| integrate | 結合；合成 |
| introduce | 介紹；引入；採用 |
| invent | 創造；編造；發明 |
| originate | 開始；發起；源自 |
| perform | 表現；執行；演出 |
| plan | 計畫；布署 |
| revitalize | 恢復生機；復活 |
| shape | 演變；進展；發展；塑造 |

表 9-4　財務能力類的 Action Verbs

| administer | 管理；治理；執行 |
| --- | --- |
| allocate | 分派；分配 |
| analyze | 分析；檢討 |
| appraise | 評價；估價；鑑定 |

| audit | 稽查 |
| --- | --- |
| balance | 平衡 |
| budget | 預算；編列預算；安排 |
| calculate | 計算；估計 |
| compute | 計算；估計；評價 |
| develop | 發展；發達；進步 |
| forecast | 預測；推測；預報 |
| manage | 管理；處理；維持；達成 |
| market | 銷售；上市；營銷 |
| plan | 計畫；布署 |
| project | 計畫；預測 |
| research | 研究；調查 |

## 表 9-5　輔導能力類的 Action Verbs

| assess | 徵收；課徵；估價；評價；評定 |
| --- | --- |
| assist | 幫助；援助；幫忙 |
| clarify | 澄清；淨化 |
| coach | 指導；輔導；訓練 |
| counsel | 商量；商議；建議 |
| demonstrate | 論證；證明；證實 |
| diagnose | 診斷；觀察 |
| educate | 教育；教養；訓練 |
| expedite | 加速；促進；發放；派遣 |
| facilitate | 促進；助長；幫助 |
| familiarize | 親近；通曉；熟悉 |
| guide | 導覽；教導；指揮；影響 |

| refer | 引用；引證；查詢；參照 |
|---|---|
| rehabilitate | 復健；復職；修復；翻新 |
| represent | 代表；描述；代表 |

表 9-6　管理能力類的 Action Verbs

| administer | 執行 |
|---|---|
| analyze | 分析 |
| assign | 分配 |
| attain | 獲得 |
| chair | 主持 |
| contract | 獲取；締結 |
| consolidate | 結合；加強 |
| coordinate | 協調 |
| delegate | 委派 |
| develop | 發展 |
| direct | 指導；經理；管理 |
| evaluate | 評價；評估 |
| execute | 執行；實行 |
| improve | 改善；改進 |
| increase | 增加 |
| organize | 組織 |
| oversee | 監督 |
| plan | 計畫 |
| prioritize | 排序；攫取重點 |
| produce | 生產 |
| recommend | 推薦 |

| review | 回顧 |
| --- | --- |
| schedule | 預定；安排 |
| strengthen | 增強；加強 |
| supervise | 管理；監督 |

## 表 9-7　研究能力類的 Action Verbs

| clarify | 澄清；淨化 |
| --- | --- |
| collect | 集中；收集；集成 |
| critique | 批評；批判；評論 |
| diagnose | 診斷；觀察；論斷；調查 |
| evaluate | 評價；估價 |
| examine | 檢查；審查；檢驗；審理 |
| extract | 選出；摘錄；取得；推導 |
| identify | 支持；確認；等同於 |
| inspect | 檢閱；檢查；檢驗；查看 |
| interpret | 解釋；說明；詮釋 |
| interview | 訪視；訪談；會見；接見 |
| investigate | 調查；審查；研究 |
| organize | 安排；組織 |
| review | 檢查；細察；審閱；審核；檢閱 |
| summarize | 概括；總結；概述 |
| survey | 通盤考慮；觀察；檢查；普查 |
| systematize | 系統化；體系化；組織化；秩序化 |

表 9-8　教學能力類的 **Action Verbs**

| adapt | 改編；改寫；改作 |
|---|---|
| advise | 勸告；建議 |
| clarify | 澄清；闡明；使～明晰 |
| coach | 訓練；指導 |
| communicate | 溝通；傳達；感染 |
| coordinate | 協調；整合；綜合 |
| develop | 發展；發達；進步 |
| enable | 使實現；使有效；使成為可能 |
| encourage | 鼓勵；支持；激勵 |
| evaluate | 評價；估價；估……值；定……價 |
| explaine | 說明；講解；解釋；闡明 |
| facilitate | 促進；幫助 |
| guide | 指導；支配；管理 |
| inform | 告知；通知；充滿；報告 |
| initiate | 傳授；引進；入門；啓發 |
| instruct | 教育；教養；委派 |
| persuade | 說服；勸說 |
| stimulate | 振奮；激勵；鼓舞 |

表 9-9　技術能力類的 **Action Verbs**

| assemble | 整合；聚集；召集；整理；裝配 |
|---|---|
| build | 建立 |
| calculate | 計算；依靠；打算 |
| compute | 計算；估計；評價 |
| design | 設計；計畫 |

| devise | 設計；發明；策劃 |
| --- | --- |
| engineer | 設計；監督 |
| fabricate | 製造；建造；裝配 |
| maintain | 養護；保養；維修 |
| operate | 操作；運轉；動手術 |
| overhaul | 徹底檢查；翻修；精細檢查 |
| program | 規劃；計畫；製作 |
| remodel | 改造；改型；改變 |
| repair | 修理；補救 |
| solve | 解決 |
| train | 訓練；練習；鍛鍊 |
| upgrade | 提升；升級；提高品級 |

 履歷要件解析

　　履歷表是求職時最重要有效的自我行銷工具，也是求職者獲得工作的重要關鍵，雇主依照履歷來決定是否找你面談。因此，能讓你從眾多競爭者中脫穎而出、贏得面試機會的，就是份能清楚表達你個人經歷的履歷表。

　　雇主由於時間有限，只會快速瀏覽，在數十秒內就決定要進一步聯絡還是淘汰，履歷表扮演的角色就是在這關鍵的數十秒內說服對方。你的履歷若不能很快獲得注意，就會被其他履歷所淹沒。

　　要製作一份出色的履歷表，在眾多應徵的履歷表裡爭取到面試的機會，就應把握簡潔原則。因此，不論中英履歷，篇幅都不要超出 2 頁。

　　履歷要能清楚呈現自己的專長與學歷背景，要文筆流暢，讓雇主

在短短篇幅中，發現你的特質與專長，了解你對該工作的企圖心而將你歸屬於「合於複試資格」者。

除非該職位需要你表現設計創意或電腦技巧，否則不用把履歷表做得很花俏。真正應該注意的是文法要正確、沒有錯別字、字體要合宜、段落和句子要整齊、編排要得體、格式要清爽易讀、資歷重點要突出等。

履歷中當然沒有理由加入對自己不利的資料。

有人動輒使用英文字縮寫（Acronyms），認為只要寫出來就該有人認得。事實上，即使是專家，對自己領域中的縮寫也不見得完全清楚，因此在使用縮寫時，應該至少引用一次全稱。你雖可以自己擴充縮寫詞，但必須在該縮寫詞第一次出現時，用括弧將全稱括在裡面。例如：

Retail business in Taiwan has adopted DM (Direct Mail) as advertisement media. DM has been gaining its popularity since 1980's.

但是已經為大眾所熟悉的節略字，如 laser、MBA radar、USA 等，則可以直接使用。

中英履歷都應該包含下列的必要項目：

1. 個人基本資料
2. 專長
3. 學歷
4. 經歷

本章各節將引用英文履歷為範例，把履歷表的各個部分詳細加以敘述。

## 一、基本資料（Personal Data）

個人基本資料主要包括姓名、地址、電話、e-mail 等，詳細通訊地址及聯絡電話千萬不能有誤。

如果履歷表的空間許可，個人基本資料可以寫成如下：

Naichia Yeh

1619 Gruenther Ave.,

Rockville, MD 20878

O: 240-632-9274

H: 301-331-5689

naichiayeh@hotmail.com

或：

Name: Naichia Yeh

Address: 1619 Gruenther Ave., Rockville, MD 20878

Phones: (O) 240-632-9274, (H) 301-331-5689

e-mail: naichiayeh@hotmail.com

如果履歷表的空間不多，則個人基本資料可以寫成如下：

Naichia Yeh, 1619 Gruth St., Towson, MD 20878

O: 240-632-9274, H: 301-331-5689

naichiayeh@hotmail.com

就像你不會在履歷上寫出你的生肖一樣，星座當然可以免了，小心碰上對星座的觀念特別反感的雇主，他們把「星座說」當成是愚蠢的事。

像是生日、婚姻狀況、嗜好、身高體重及照片等與工作能力無關的事項，大多可以省略。事實上，在歐美職場中，詢及年齡及婚姻狀況是無禮，甚至是侵犯隱私權的，因此雇主在徵人時，於法不能、也不會要求職者在履歷中列明該等項目。這一點在非英語國家似乎有所不同，所以在法律允許雇主限定求職者性別、年齡及婚姻狀況的社會也只好不那麼堅持，要依雇主的要求一一列出：

Name: Mr. Naichia Yeh
Address: 1619 Gruth Ave., Main, MD 20878
Phones: (O) 240-632-9274, (H) 301-331-5689
e-mail: naichiayeh@hotmail.com
Date of Birth: 1980/9/1
Marital Status: Single

當然，雇主若要求附上照片，你也只好遵照規定。請注意提供的照片中應儀容整潔且服裝正式，社會新鮮人則可使用學生照。

某些個人資料（如嗜好等）若對工作有幫助則加上無妨，例如，愛好運動及旅行也許可以表示適合旅行或外調的工作。

二、專長（Qualification Summary）

敘述專長時，可依其與職務相關性的高低順序，採用要點方式條列和說明，例如：

Qualification Summary
1. Computerized Environmental Hazard Ranking, Rating, and Scoring System Development
2. Computerized Environmental Decision Support System Design and Development
3. Environmental Data Retrieval System Design and Development

　　當然也可以適當使用大寫、**黑體**、斜體或底線等，突顯重要的、能吸引雇主的資歷：

QUALIFICATION SUMMARY:
1. <u>Environmental Audit and Regulatory Compliance Assessment</u>
2. *Hazard Communication, Environmental Risk Management, and Emergency Response Planning*
3. Hazardous Waste Minimization and Waste Management Plan Implementation
4. Preliminary Assessment/Site Investigation, Remedial Investigation/Feasibility Study, and Risk Assessment

　　如果篇幅允許，在專長項目之前，可用一段前言來強調自己的賣點，注意陳述要簡潔，句子不要太長：

Qualification Summary:

Has extensive experience in marketing and developing computerized environmental system and projects. Specific qualification includes:

1. Software system programming, data retrieval and analysis, user interface design and development

2. Computer-aided modeling, Geographical Information System (GIS) application

3. Environmental (water, air, and soil) models, database, multimedia environmental information management systems, user training, technical support

　　現今的行業無一不需用到電腦，因此你若有電腦應用方面的知識，也可將之列入專長項目裡。

　　若所應徵的是電腦方面的工作，不妨在履歷表上另開「電腦專業能力」部分如下：

COMPUTER EXPERTISE：

1. Hardware: IBM-PC family, MacIntosh family, IBM 360/370, IBM 4341/4381, VAX etc.

2. OS: MS-DOS, Windows, MacIntosh System, CPM, Music, Wylbur, VMS, etc.

3. DBMS: DBase family, FoxBase, FoxPro, Visual FoxPro, Clipper, etc.

4. Languages: BASIC, PL/1, FORTRAN, Pascal, HyperTalk (for HyperCard), OpenScript (for ToolBook and Multimedia Tool-Book), etc.

　　當然，如果你有特殊的語文能力或機械操作能力，也可以列在專長項下。

## 三、經歷（Work Experience）

　　若有 2 年以上的工作經驗，可把經歷列在學歷前面，而若是剛畢業不久，沒有重要的工作經驗則最好先列學歷，再列經歷。

　　經歷部分應由最近的工作開始，以反時間順序詳實列出公司名稱、職務、任職起迄，並且扼要敘述工作內容，如能附上優良工作成績的紀錄或說明更好。

### ㈠工作經驗描述要領

　　一般而言，雇主希望履歷表上有比較具體的工作經驗描述，藉以判斷求職者是否是他所需要的人才，先舉一個模糊的例子：

　　Specializes in system analysis, system design, and network administration

　　具體的寫法應該是：

Computer Skills
Database: Oracle, INFORMIX, ROGRASS, DBase, DBMS

System: Unix, MS-DOS, Windows NT, WWW server

Language: C, C$^{++}$, COBOL, Visual Basic, Java, FoxPro, Clipper, Oracle Developer 2000

再來一個模糊的例子：

具備很好的分析能力。

必須要這樣說才算具體：

曾於某公司的投資個案中，及時偵查出財務報表中資金的異常流向，因而撤回原訂的投資案，該公司於半年後發生資金掏空案。

有數字佐證的量化敘述，比非量化的泛泛說法要具體，下面的幾個實例應該可以說明模糊與具體的分別：

模糊：能夠於極短的時間內完成一份內容可觀的企劃案。
具體：能夠在三小時內製作完成一份長達二十頁的企劃案。
模糊：負責督導東南亞地區的業務及銷售。
具體：獨力規劃和訓練一個由十五名代表所組成的業務團，以
　　　一系列的行銷活動在三個月內成功贏得東南亞地區60%
　　　的主要通路。
模糊：任職期間每年都能順利達成業績目標。

具體：在任職三年內取得公司最主要客戶之代理權，並且分別
達成 112%、121% 及 135% 的營業目標。

模糊：負責系統安裝、維護、除錯及售後服務。

具體：有效運用問題解決模式，將現場系統障礙排除能力在
1 年內由 23% 提升至 55%；客戶滿意度由 62% 升高至
92%。

　　如果實在經驗有限，缺乏量化的事實作基礎，也可以換個方法自
我推銷，這時就不妨「利用」模糊。因為在重視包裝的社會裡，對自
己的資格稍加美化也是無可厚非的自我推銷做法。

　　例如，工讀生做的工作是：

端茶和代訂便當
影印及收發文件
發放傳單

可以美其名為：

負責處理日常庶務
負責檔案管理
推廣業務、接觸人群及自信訓練

對於較低微的工作職稱，也可適度加以「調整」，例如：

　　工友或工讀生
　　行政助理


不妨轉稱爲：


　　庶務助理
　　行政秘書


　　社團裡的幹部名稱及學生活動裡的幹事人員，若是用上「執行長」，「總務長」、「公關主任」等比較「高級」的名詞，多少有點取法乎上的鼓舞作用，也有突顯業務與衆不同的功能，是種向高處瞄準的積極思想。與其把這類的做法說成是吹噓或是自我膨脹，不如想像你是在自我期許，因爲一般人能做的事情絕對比他們目前所做的更多、更好。

　　只要有信心，求職時可以把自己目前的能力提高個 20%、30% 來描述，重要的是自己願意繼續充實，只要在獲得該項工作前，達成所寫下的目標，就算是對自己誠實。

(二) 經歷豐富者的寫法

　　一般的履歷表宜簡單明瞭，免得重要的訊息被太長的敘述淹沒。因此，寫作履歷與其長篇大論，不如簡要易讀。

　　但是，如果謀求的是較高的職位，就不免需要較有內涵的資歷，資深者的履歷表必然會比較充實，篇幅難免會比較多，節制篇幅的好辦法是用不同的字體（如粗體、*斜體*、加底線、加大字體甚或前述*多種方式的混用*等的方法）來突顯主題，英文大寫（ALL CAPS）及重

點列舉（Bulletize）的方式等都會有用。

　　一般來說，雇主對應徵者最近的工作經驗較有興趣，可以適度多加介紹：

---

**WORK EXPERIENCE:**

**General Science Corp., Sr, Engineer, 2002 - Present**

1. **Conducted air and water pollution modeling to facilitate Toxic chemical de-listing petition evaluation.**

2. **Performed parameter sensitivity analysis for comparison and verification of modeling results.**

3. Conducted implementation and technical support of Environmental models including air, soil, and groundwater. Tasks included assisting contract management, customer relations, user training, software quality assurance and quality control, user's manual development, software testing, etc.

InfoTech Inc., Senior Scientist, 1995 - 2001

1. **Conducted a feasibility study project on the establishment of Asian Environmental Business Information Exchange/ Clearinghouse.**

2. Managed system analysis and computerized management services for various projects. Responsibilities include market investigation, graphical user interface (GUI) integration, real-time data acquisition via virtual instrument, and geographical information system (GIS) R&D.

## The Univ. of Texas, Teaching Assistant, 1995 - 1998

1. Taught undergraduate business mathematics and computer programming

2. Instructed graduate students on their research projects

## Peitou High School, Science Teacher, 1991 - 1995

1. Conducted physical science curriculum design and development

2. Taught high school physics, mathematics, computer, energy, and environmental courses

　　履歷表必須瞄準目標，涵蓋所有的個人行銷資訊，至於涵蓋期間，通常只要最近的十年經歷就夠了。如果經歷實在太長，那就不妨省略下列的項目：

1. 類似工作的職務說明
2. 低階工作職稱
3. 不相關的工作經驗

依照這些省略原則，我們可以把上例刪改如下：

## WORK EXPERIENCE:

## General Science Corp., Sr, Engineer, 2002 - Present

1. Conducted air and water pollution modeling to facilitate toxic chemical de-listing petition evaluation.

2. Performed parameter sensitivity analysis for comparison and verification of modeling results.

3. Conducted implementation and technical support Environmental models including air, soil, and groundwater.

**InfoTech Inc., Senior Scientist, 1995 - 2001**

**1. Conducted a feasibility study project on the establishment of Asian Business Information Clearinghouse.**

**2. Managed system analysis market investigation, graphical user interface (GUI) integration, real-time data acquisition via virtual instrument, and geographical information system (GIS) R&D.**

當然，如果覺得過去數十年的經驗對這份工作都有幫助，那就不妨都列進去。

㈢ 職場新鮮人的寫法

剛踏入社會的職場新鮮人若無太多經歷背景，則可把重點放在下列事項：

1. 學校社團領導人的角色
2. 課外活動的成績
3. 義工或兼職工作
4. 實習工作或學徒經驗
5. 社會服務
6. 學校或社會的社團經驗
7. 旅行
8. 語言能力
9. 電腦專長

熱誠、成熟、精力旺盛等正面的個人特質，在工作上都會有加分

效果，在校的主修課程若成績極佳，不妨加列學業成績。下面的例子
提供一些水準以上的示範：

1. Excelled in scheduling, prioritization, and time management
   via maintaining a part-time job while attending school full-
   time.
2. Commended for superb research and project reports by profes-
   sors and internship supervisor.
3. Excelled in communication skills via numerous class presenta-
   tions, daily communication with students and staff to clarify
   requests.
4. Provided administrative information and resolved problems,
   on regular basis, in a professional manner as an elected assistant in the Office of Student Affairs three terms in a roll.
5. Developed great teamwork skills via intensive team projects
   required by cause work.
6. Maintained excellent coursework at School and prior internship with a 9 GPA.

　　若能透過自傳具體陳述打工經驗、社團研習活動、參與過的學
校研究或專案計畫等，也會有幫助。那些經歷多少能顯示些個人特
質，如志趣、合群性、領導力、成熟度等，可作為雇主的參考。

## 四、學歷（Education）

> BA, Information Management (minor in Education), National Taiwan Normal University, Taiwan, 2000

　　有些雇主重視證照，若能在履歷表中列舉獲得的證照（例如電腦相關認證），甚至將證照的影本或圖檔附上，則可能會吸引審核者的興趣。

　　一般可以把所考取的證照列在學歷項下：

> EDUCATION AND CERTIFICATION:
> 1. M.S. Physics (minor in computer sciences), Myown University, Tainan, 1998
> 2. B.S. Physics, Taiwan Normal Univ., Taipei, 1995
> 3. CHMM. (Certified Hazardous Material Manager), HazMat Management Institute, 2000

　　此外，政府或學校提供的專業認證（例如就業專班、學分班的結業證書等），技術類及語文類（TOEFL 或全民英檢）的證照等，同樣能產生加分效果。

## 五、其他項目

　　履歷表中除應包含個人基本資料、專長、學歷及經歷等四要項外，還有應徵職務、介紹人名單及期望待遇等項，依序說明於下。

## (一) 應徵職務（**Position Desired**）

　　廣告上有時會要求應徵者列明屬意的職稱，此時就要把「應徵職務」這一項加在履歷表上，如果該廣告內有列明職稱，你只要在履歷上照錄職務名稱即可，例如：

1. Product Manager
2. Administrative Assistant
3. Sales Representative
4. Software Engineer

　　當然也可以寫得詳盡具體一點，有些過時的履歷會加寫些流俗的贅字，例如：

1. An elementary school media consultant position <u>with growth potential</u>
2. <u>A challenging position of</u> a program analyst with emphasis in real-time transaction systems

　　類似 challenging position 或 growth potential 並不具有實質的意義，大可刪去：

1. An elementary school media consultant position
2. A program analyst position with emphasis in real-time transaction systems

　　若不是由求才廣告得來的資訊，由於無法確定雇主對該職務的稱呼，通常只好寫一個籠統的職位，而不寫也沒太大關係。

## ㈡ 介紹人（References）

　　雇主多在決定要聘用當事人時，才會要求提供介紹人或推薦人名單，因此，若應徵條件中沒有要求，履歷表上就不用自動註明推薦人的名字，只要在履歷表之末寫上：

References available upon request

就算得體了。

　　而若雇主有要求，則可在履歷表末列出一兩位過去主管或師長的聯絡方式或電話。例如：

References:
1. Dr. Gary Yeh, Professor, University of Texas, Austin, TX., garynyeh@hotmail.com, 240-207-4625
2. Mr. Sam Liu, President, ITI, Inc., Rockville, MD, samliu@itimd.com, 301-268-2079

　　有些雇主會要求職者列出三個推薦人，請他們佐證求職者的專長背景與能力，為了因應這類不同的需要，你可以先把推薦人的資料準備在另外的紙上，以備不時之需。

　　推薦人表中可以列明你與推薦者的關係，如有必要，也可以註明雇主適合去電的時間，例如：

References:

1. Mr. Frank Lee

   President, Lee Hardware

   123 Fifth Street, Tupleo, MS 34567

   e-mail: lee@lhardware.com

   Work hone: (__) ____-_____, Monday through Thursday

   Relationship: Supervisor at ABC Corporation

2. Dr. Ellie Sand

   Professor, Myown University

   456 West Seventh Street, Racine, WI 65432

   e-mail: esend @myown.edu

   Work phone: (__) ____-_____, weekdays before noon

   Relationship: Academic Advisor

3. Ms. Martha Pahlavi

   Director, ABC Corporation

   1010 Balsam Boulevard, Dallas, TX 76543

   e-mail: pahla @abcoil.com

   Work Phone: (__) ____-_____

   Relationship: Served Together as a Project Leader

　　列出任何人的名字及通訊方式之前，一定要取得他們的允可，徵求同意。一方面是尊重他人隱私權的禮貌做法，一方面也是請他們有心理準備，免得雇主來詢時不知到底是怎麼回事。

　　請記住，非經首肯，千萬不要列入推薦人家裡的電話，這是對私領域的尊重。

## (三) 期望待遇（Preferred Salary）

期望待遇一向是敏感的議題，求職者最好先研究一般的薪資水準，再寫出公司與自己都可接受的彈性額度，還好這年頭資訊發達，各行業和年資的薪水標準都可以在網路上找到。

還有，你也可以權衡一下自己對該職位需求的迫切程度，已經有工作但想跳高薪水的人，當然可以待價而沽，不用太屈就於不滿意的薪水，沒有正式工作的人也許就只好委屈點，遷就一下雇主願意給付的代價。

歐美薪水是以年薪表示，東方國家則多以月薪表示。在英文文件中，錢幣符號「$」多代表美金，如是台幣則應註以「NT$」。以下是期望待遇欄中可以使用的格式：

1. Negotiable（面議）
2. High $20s（近 30,000）
3. Upper $30s（37,000 － 40,000 萬之間）
4. Mid $40s（44,000 － 46,000 之間）
5. Lower $50s（50,000 － 53,000 之間）
6. Low $50s（50,000 出頭）
7. NT$50,000/month, preferably（台幣月薪）
8. Preferred salary: $60,000 per annum（美金年薪）

期望（Preferred）比要求（Required）較有協議空間，議薪時只要合乎常理，不用過於保守，因為要求的薪水比平均值低不見得會增加多少錄取機會，如無把握則可「依公司規定」。

## 完整履歷的範例

　　履歷表紙張以白色或米色為佳，避免使用彩色紙張，大小以 A4 或 8.5"x11" 為原則，段落之間保持適當距離，能給讀者喘息的空間，是迎合讀者心理的做法。同時用紙的厚度及列印品質也要講究，太薄的紙張會影響整個呈現的品質。列印時應使用雷射印表機，點陣式印表機品質較粗糙，應該避免。

　　履歷表固然以內涵為重，但外觀也能有決定性的效果，因為履歷表的編排如果過於雜亂，雇主也許還會對求職者的印象不佳。所以資歷固然應該詳盡說明，但也不要在一兩張紙上寫得密密麻麻。

　　在正式寄發履歷表出去之前，應踏實做好最後確認工作，除了自己校對，還應請他人校閱。

　　郵寄時應選擇合適大小的封套，不要裝訂太複雜，若公司要求在信封外註明應徵項目或其他資料（如請註明「人事部」收）等，也記得要一併註明，以方便公司分類。若以郵寄的方式寄出，記得貼足郵票以免被退回，增加往返郵寄時間或給予公司粗心的印象。

　　以下舉出一些完整履歷的例子，其格式、內容及撰寫的方式都足以供謀職者參考。

　　圖 9-1 和 9-2 的單頁英文履歷適用於資歷較淺者，圖 9-3 和 9-4 分別為中文與英文，都涵蓋了 2 頁的篇幅，是相當資深者的履歷。

　　有些求職者希望雇主能為其前來申請工作及面試約談一事保密，如果你也有這樣的考量，可以像圖 9-1 一樣，在履歷表首行加上下述的字樣：

　　Confidential Resume of...

**CONFIDENTIAL Resume of** _____

Address: _____

Phone:    H: _____    O: _____

**Objective:**

A Programming Analyst position with emphasis in real-time transactional systems

**Qualifications:**

Languages:     C, Pascal, BASIC, COBAL, C++

Hardware:     IBM, Macintosh, VAX, NOVA

Software:     DB2, BTrieve, Oracle

**Work Experience:**

Programmer Analyst, ABC, Inc., Austin, TX (2005 － )

1. Maintained software publication programs.

2. Developed application software.

3. Provided technical backup.

4. Coordinated and designed system resulting in increased efficiency.

Software Specialist, ITI, Inc., Taipei, ROC (2002 － 2004)

1. Scheduled, supervised and trained tech support staff.

2. Used analytical and writing skills to communicate with clients, employees and management in a professional manner.

3. Certified and tested new software programs.

**Education:**

1. M.S. Computer Science, University of Maryland, College Park, Maryland, 2002

2. B.S., Information Technology, Myown Universty, Taipei, Taiwan 2000

圖 9-1　資淺者之英文履歷範例（密件）

Mr. _____

H: _____ O: _____ , nyeh@hotmail.com

**Objective:**

An elementary school media and reading consultant

**Achievements**

1. Developed and implemented individualized educational plans.

2. Tutored underachievers in remedial reading with a better than 60% improvement rate.

3. Introduced successfully audiovisual techniques in county wide reading program.

4. Cited as Teacher of the Year.

**Work Experience:**

Media Teacher, Brook Elementary, Dallas, TX, 2004 －

1. Prepared outlines for daily and monthly course of study for grades 1-

2. Lectured and demonstrated with audiovisual teaching aids.

Teacher, Alachua Elementary, Gainesville, FL, 2002-2003

1. Counseled parents and directed them into remedial action for specific cognitive or emotional problems of children.

2. Counseled and directed children with reading difficulties to achieve higher reading comprehension.

**Education:**

BA, Education, Univ. of Florida, Gainesville, FL, 2002

圖 9-2　資淺者之英文履歷範例（一般）

Dr._____

**1230 Main St., Gaith, MD 20878**

**(301) 208-0276 e-mail: gyeh@hotmail.com**

**QUALIFICATION SUMMARY**

Has 16 years of combined academic and technical experience in supplying technology based solutions to environmental management for government agencies and private industry. Specific qualification includes:

1. Industrial safety, environmental audit, and compliance process development
2. Science, engineering, and management software system user-interface design and development
3. Statistical air quality analysis, numerical simulations, and sensitivity analysis capabilities on different science and engineering systems

**EMPLOYMENT HISTORY**

**InDyne, Inc., Director of Technology, 2003-Present**

1. Provided the concept and technology of electronic manual to National Highway Safety. The marketing effort resulted in the design and development of Electronic Data Collection and Coding Manual for National Accident Sampling.
2. Designed and developed a graphic oriented system that integrated NASA cost, project, personnel, and space science and technology data uses multimedia concept and technology.

1/2

圖 9-3　資深者之英文履歷範例

……接上頁

### International Tech Inc., Sr. Scientist, 1999 - 2002

1. Conducted a feasibility study project on the establishment of Asian Environmental Business Information Exchange/Clearinghouse.
2. Managed system analysis and computerized services for real-time data acquisition via virtual instrument, and geographical information system (GIS) R&D.

### AEPCO, Inc., Senior Scientist, 1995 - 1998

1. Conducted corporate overseas marketing activities in China, Taiwan, and other Asia countries. Provided technical presentation and organized seminars/conferences to support corporate business development effort in Asian environmental Information management market.
2. Lead a U$5 million contract in developing the world's first large-scale computerized Chinese language multi-media chemical emergency response system from concept to implementation. This system includes a state of the art air plume model and has been used by over 2,500 chemical plants to provide decision support functions for monitoring industrial emergency responses.

### EDUCATION AND CERTIFICATION

1. Ph.D., Physics, Univ. of Texas, Mytown, TX., 1994
2. M.S. Computer Science, Myown Univ., Tainan, 1991
3. B.S. Physics, Taiwan Normal Univ., Taipei, 1989
4. Certified Hazardous Material Manager, 2000

2/2

圖 9-3（續）

簡歷
_____

(　　)　-　　　　，e-mail:abcd@hotmail.com

資格與專長：

　1. 十餘年學界及業界之管理決策支援、業務開發、教育訓練及研究發展實務

　2. 精於數理分析、電腦軟體設計及研發各類資訊系統和電子出版品

　3. 長於中英文寫作，具有紮實專案管理及技術背景及豐富教學經驗。

學歷：

美國_____大學環境管理博士 (_____ University)

美國_____大學資訊管理碩士 (University of _____)

國立_____大學物理學士（輔修教育）

經歷：

美國 InfoDyn, Inc., 科技部主任，1999 迄今

　1. 主持臺灣及美國環保署之「亞洲環境資訊中心在台設置可行性」研究計畫。

　2. 主持公司環保業務在臺灣及大陸拓展之計畫。

　3. 主持整體性電腦環境管理系統之發展、市場調查及相關技術開發細節。

　4. 主持美國環保署之 Graphical Environmental Models 專案計畫中空氣、土壤及水汙染模式之研究、技術支援與使用介面整合。

1/1

**圖 9-4　資深者之中文履歷範例**

……接上頁

美國 Abc, Inc., 資深工程師，1995 － 1999

1. 利用多媒體資訊系統完成臺灣環保署委託之電腦化工業安全系統之設計與開發，此系統為第一套中文環境決策支援系統，計使用於約 1500 家相關產業。
2. 主持美軍基地整建計畫環境評估、勘驗與補救可行性研究等專案。
3. 編撰美國海防環保法規技術指導手冊以供環保法規執行成效考察之用。
4. *參與美軍危害廢棄物減廢（Hazardous Waste minimization）計畫成效考察及資料蒐證。*
5. 主持美國陸軍之環保考察專案，稽查各重要軍事基地環保法規執行之成效。

Dynamac Corp., 研究員（Scientist），1993 － 1994

1. 完成一電腦化害度評估系統 (Hazard Ranking System)，此系統成為數百危害廢料場之評估標準，經廣泛使用於各類勘驗報告內，為美國戰略總署危害物料技術中心勘驗報告之範本。
2. 撰寫各類技術及法規之研究企劃，提供危害物料管理之法令諮商及技術指導。
3. *主持美國空防部危害廢料處置場所之勘驗及調查，多氯聯苯、石化廢料、重金屬及石綿建材等汙染之蒐證、取樣、分析、研判與報告。工作地區遍及美國空防部所轄各州空防部隊及阿拉斯加空防司令部所轄單位。*

重要著作：

1. 《心、靈與意識》，商務印書館，2005/3
2. 《中英論文寫作綱要》，五南出版公司，2005/1
   《知識管理》全華科技圖書公司，2004/3

2/2

圖 9-4（續）

　　運用網站求職和以電子郵件寄送履歷表已屬平常，若是以 e-mail 的方式寄發履歷表，最好在郵件的標題中直接註明這是一封履歷郵件（可寫出姓名與要應徵的職務）。在正式寄出之前，可以先試寄一份給自己，看看附加在郵件中的檔案是否可以順利開啓，確認無誤後再寄給對方。

　　又，求職文件的內文要莊重得體，千萬不要把履歷表弄得太花俏，表情符號或動畫人物在非正式書信中雖然有時很討喜，但並不會為你的履歷加分，反而很有弄巧成拙的可能。

　　對於直接線上登錄履歷表，可先在文書處理軟體中擬好內容，再轉貼到履歷登錄頁面的欄位中，才不會每次都要重新打字。

# 第十章
# 自傳與求職函件

本章提綱

◎自傳（Autobiography）

◎相關函件

求職與應徵函（Cover Letter）

推薦函（Recommendation）

謝函（Thank-you Letter）

接受與婉謝函

## 自傳（Autobiography）

在美國求職甚少需要寫作自傳，但在海峽兩岸，雇主要求履歷附上自傳似乎已成為常態，因此我們也只好隨俗。

自傳是個人背景以及學、經歷、專長的綜合說明，避免過於艱深的語句，宜摘要補充履歷之不足，多強調自己的優點與想法，英文自傳的字數在 300 字上下即可。由於自傳描述的是自己，因此使用較多的 *I* 乃是無可厚非的事。

以下且來分解一篇英文自傳的佳作，第一段：

> Born on _____ in _____, I am married with three children. I am personable in nature; like to make friends; enjoy cooking, hiking, and traveling.

　　自傳的首段簡單帶過了家庭狀況。除非你的家庭背景對公司僱用你與否有很大的影響（例如工商業背景良好，人脈雄厚，能為公司帶來業務），不然家庭背景不須寫太多。

　　本段並且大致描述了自己的興趣和人格特質（Personal Trait），服務型的公司需要積極拓展業務，親切的態度是多數服務業對員工的第一要求，而研究性的工作可能要求員工專注和關心細節，如果你的特質與工作性質相合，就可以在這一部分加以敘述。

　　再來看第二、三、四段：

　　After graduated from _____ University in Taiwan with a BS degree in physics, I started as a research assistant in _____ Company. My first job was a project about image/graphic technology research. The job has made me fell in love with research work related to computer technologies. As a result, I decided to go abroad to pursue an advanced degree in computer engineering.

　　19__, I went to _____ University in _____ to study _____. During the course of my Ph.D. study, I published three important papers at three accredited journals in the field of _____. These papers later helped me secure a position in _____ lab where I met my wife, who is also from Taiwan. We dated for about two years. Then she and I decided to get married and moved back to Taiwan where each of us had a college teaching position in waiting.

　　In _____, I joined _____ Company in Taipei as a head scientist. Lead several R&D projects, including the study of a man-machine interface, development of a desktop publishing system, and enhancement of the Chinese MS-Windows. After five years of intensive technology-based work, I feel it is the time for me to move on to a more management-oriented career. This decision has lead me to contact your company, _____ .

這一部分敘述了求學和工作的歷程，這些條目雖然在履歷中已經列明，但是在此仍可將限於篇幅而無法列在履歷表中的內容稍加敘述。如果工作經歷與所求的職位有關，則不要忘記在此時加強描述。

接著是末段：

My strong research background and knowledge of hi-tech product life cycle analysis will help you cut down product development cost and shorten the development cycle. I am familiar with both Chinese and US cultures, fluent in four languages that include English, Mandarin Chinese, Cantonese, and Taiwanese. Such culture background and language proficiency will certainly help you develop business in Chinese and English speaking societies, which happen to be the biggest marketplaces in the world for the years to come.

末段說明了自己的強項，以及那些強項可以為公司的業務帶來的好處，重點部分用了底標加以強調，有提醒讀者的效果。

下面是一篇出色的中文自傳，以第三人稱寫作，文字流暢易讀，用 500 字的篇幅將生涯中比較重要的事件一一簡要列出，黑體字部分突顯的是與所應徵職位有關的專業，最後一段文字敘述了自己的喜

好，富有人情味，足以列為自傳中的範本：

　　　葉＿＿＿＿於＿＿＿＿年在＿＿＿＿大學物理系畢業，兩年後負笈旅美，於＿＿＿＿大學取得＿＿＿＿博士學位。學成，先後任職於數家工程顧問公司，從事與環境管理、工程專案管理及資訊管理方面的工作。

　　居美十餘年間，葉氏曾主持中美環保署合作計畫數年，為臺灣製作及引進毒性物質災害防救模擬模式、毒化物中文資料查詢系統之毒理資料庫與列管毒化物資料庫，以及毒災聯防體系氣體擴散模擬模式決策支援系統等。他時時因應公務需要，回臺掌理美國母公司之在臺業務，以及公司之工程業務在臺灣及大陸拓展之規劃，並曾主持臺灣及美國環保署之「亞洲環境資訊中心在臺設置可行性」研究計畫，於生活環境及工作經驗中養成豐厚的國際觀。

　　葉＿＿＿＿先生在 2001 年 9 月返臺定居，同年 10 月應經濟部工業局之邀，在工業安全衛生技術輔導會上，以美商＿＿＿＿＿＿公司負責人之身分發表專題演講，將＿＿＿＿監控系統網路化的觀念落實於臺灣工業界，隨後任教於南部及中部數所院校迄今。

　　葉君學過物理，做過管理，而情有所鍾於心理，他擅長中英寫作，教學期間，著述不輟，於＿＿＿＿年至今＿＿＿＿年餘，共發表＿＿＿＿、＿＿＿＿及＿＿＿＿之專著二十種，共百餘萬言。他性喜交友，以直心待人，與同仁相處融洽，並深受學生歡迎及敬重。閒暇時自炊自食，除讀書寫作之外，以騎單車下鄉漫遊於田野之間及探索老式城鄉建築為樂。

　　雇主對你的生平記事沒有興趣，不要著墨太多，重要的是該強調自己的專業素養能為公司帶來多少利益，因此，在自傳中可以簡單交代自己與工作相關的的興趣或特殊技能，此點常可增加審核者的印像，並可成為面談時的話題。

　　最後，千萬不要再用老掉牙的「余名○ XX……」來做自傳的開頭，這種開頭暴露了寫作者的迂腐和沒有創意，連抄襲都不懂得要找個好的範本。

## 相關函件

　　與求職有關的的函件大致上包括了
1. 應徵函（Cover Letter）
2. 推薦函（Letter of Recommendation）
3. 謝函（Thank-you Note）
分別敘述在下面三個小節。

### 一、求職與應徵函（Cover Letter）

　　在美國謀職甚少需要寫作自傳，但應徵函倒是不能忽略的。

　　Cover Letter 與 Resume 有互補作用，讓求職訊息更完整。一封好的應徵函要能針對雇主的人才需求適時推薦自己，爭取面試的機會。

　　先來看看下面幾個應徵與求職信的開場白，其中第一、二兩句適用於回應求才廣告的應徵信；第三句雖是以 *I* 開頭，但是不失為得體的佳句，第四、五兩句適用於不確定對方是否有求才需要時的求職信，第六句則適用於即將畢業的大學生朝職場伸出觸角的第一封謀職信。

1. It is a pleasure to learn about your opening for a _____ position.
2. Please accept this letter and resume as an application for the position of _____ advertised in _____ .
3. I would like the opportunity to put my education and experience to work for your company.
4. Does your company anticipate the need for an _____ ?
5. As a recent graduate of _____ University with a Degree in _____ , I am eager to apply my skills and experience in the field of _____ .
6. I am attending _____ University and plan to complete my Degree in July 2016.

　　再來看看幾個應徵與求職信的結尾方式，其中第一句純粹以道謝收尾，第二、三句都在試探面談的可能性，第四句則言明會去電探詢面談的機會，是個主動出擊的例子。

1. Thank you for taking the time to review my resume.
2. It will be a pleasure to speak with you in person about my qualifications and this position.
3. Look forward an opportunity to meet with you to discuss the position and my qualifications.
4. I will call you next week to see when your schedule might permit a meeting (interview).

　　有時候 *I* 真的是無法避免的，下面幾個應徵與求職信的結尾例子就全都用上了 *I* 開頭，但也都很得體。

1. I am eager to learn more about the position and describe my qualifications to you.
2. I would be very pleased to discuss the position and my qualifications further with you.
3. I look forward to having the opportunity to meet with you to discuss the position and my qualifications.
4. I would appreciate the opportunity to meet with you personally to discuss your needs and how I could contribute to your company's success.
5. I'll call you next week to see about the possibility of arranging an interview (meeting).
6. I look forward to your reply.
7. I look forward to hearing from you so that we can arrange a meeting.

　　以下介紹一些不同水準的完整應徵函，供讀者當作寫作時參考。

　　左欄這封是寫給職業介紹所請求代為謀職的信，雖然沒有文法上的錯誤，但是文意上不太順暢。原來英語國家的人也不見得寫得出順暢的英文，回頭來看看許多華語國家的人，要他們用中文寫作也可能辭不達意。

　　把左欄的信改成右欄那樣，看看有何差別：

| 原信 | 修訂後 |
|---|---|
| Dear Mr. _____ : | Dear Mr. _____ : |
| I am writing to you because I am seeking new opportunities for career advancement in _____ field. Perhaps one of your current clients may have some interest in my background. | I am seeking for career advancement in the field of _____ and hope some of your clients may have interest in my background. |
| My credentials include a _____ degree and _____ years of experience in _____, during which I have been involved in the full complement of _____ operations and have effectively progressed through various levels of employment. | With a _____ degree and _____ years of experience in the field, I have progressed through various levels of employment as stated in the enclosed resume. |
| Should you be aware of a suitable opportunity with one of your client organizations, I would appreciate hearing from you. I can be reached during the day at _____. | Would you contact me at _____ if any of your client organizations come up with a suitable opportunity? |
| Thank you. | Thank you for your assistance. |
| | Yours truly, |

　　比較負責的獵頭公司（head hunter）會為雇主篩選履歷及應徵函件，其作用有如雇主的人事單位，因此，寫給這類公司的求職資料也需要維持一定的水準。

　　左欄這封是直接寄給雇主的應徵信，在文意上表達得稍微好了些，但還是有改進的餘地，將原信簡化成右欄那樣就好多了：

| 原信 | 修訂後 |
|---|---|
| Dear Mr. _____ : | Dear Mr. _____ : |
| Enclosed please find my resume in response to your recent advertisement in the _____ edition of _____ newspaper for a _____ position. This position sounds exciting and I would welcome the opportunity to discuss it further with you. | It is a pleasure to enclose my resume in response to your recent ad in the _____ dated _____ for a _____ position. |
| Should you also agree that my background is a good match for your requirements, I would welcome the opportunity to meet with you to further explore this excellent opportunity. I feel confident that I can provide the kind of leadership that your are seeking for your company's total quality effort. | Please allow me an opportunity to meet with you for a formal interview if my background meets with your requirements. You will find that I have the kind of leadership you are looking for. You can reach me during the day at _____. |
| I can be reached during the day at (__) _____ - _____ . | |
| Thank you for your consideration, and I look forward to hearing from you. | Thank you for your consideration, and look forward to hearing from you. |
| Yours truly, | Yours truly, |

　　有些求職者希望雇主能為其求職行動保密，寫下面這封信就是這樣的一位求職者。對於不善以英文表達的人士，寫得越多越容易暴露自己不善於文字表達的弱點，這時候，「因陋就簡」反而是較好的策略，下面這封信只用了不到三分之二的篇幅，也已經足夠表達原信的意思：

| 原信 | 修訂後 |
|---|---|
| Dear Mr. _____ :<br><br>I have recently decided to effect a career change and am now confidentially exploring employment opportunities at the senior level in general management. This search is being conducted on a highly confidential basis, since my current company is not yet aware of this decision.<br><br>I am a graduate of the _____ University with a _____ degree in _____ .<br><br>Currently I hold a management position in _____ Company. In this capacity, I report directly to the President of our company and am responsible for maintaining exceptional levels of quality within our production facility.<br><br>The prospects for growth within _____ Company are not particularly encouraging for the foreseeable future. Also, there is no major business expansion planned over the next several years. At this point in my life, I feel I must move on if I am going to realize my longer-term career goals.<br><br>Mr. _____ , as a leading executive of one of this industry's most profitable corporations, should you be aware of a suitable management opportunity, I would very much appreciate hearing from you.<br><br>I can be reached on a confidential basis at _____ . | Dear Mr. _____ :<br><br>I am making a career change and confidentially exploring employment opportunities at the senior level in general management.<br><br>As you will find from the enclose resume, I hold a management position in _____ , reporting directly to the company president.<br><br>While I have been maintaining exceptional quality within the company's facility, the prospects for growth within this company are not encouraging since there is no major business expansion planned over the next few years. You can see why I must move on for a longer-term career goal.<br><br>If you are aware of a suitable management opportunity, would you contact me confidentially at ( )_____ - _____ ? |

| | |
|---|---|
| Thank you for your assistance in this matter.<br>Sincerely, | Thank you for your assist-ance in this matter.<br>Yours truly, |

左欄這封信的語氣不是很柔和，就像一個陌生人來請你幫忙，可是臉上卻不帶笑容。改成像右欄那樣之後，語氣柔軟了許多。

| 原信 | 修訂後 |
|---|---|
| Dear Sir: | Dear Personnel Manager: |
| I am responding to your advertisement on _____ and applying for the position of _____. Enclosed you will find my resume, which will give you all the applicable information about my qualifications for this position. Please contact to get more information or to set up an interview.<br>I look forward to hearing from you.<br>Sincerely, | I am responding to your advertise-ment on _____ for the position of __ ____. The resume, as enclosed, will give you the applicable information about my qualifications. Would you contact me if you need further information or are ready to set up an interview?<br>Sincerely, |

既然有求於人，何妨多禮一些？

以下舉兩個適合初至中級求職者的 cover letter 範例，第一封是向已經公告職缺的公司應徵：

Dear Ms. _____:

I graduated from _____ University with an MBA, concentration in _____. I was very excited to learn of your opening for an __ ____ and am enclosing my resume in application for the position.

Your description of the _____ position looks like an excellent match for my qualifications. I offer:

1. Strong Organization Skills - Attending school full-time while maintaining a part-time job has helped me become an expert in scheduling, time management, prioritization and efficiency.

2. Excellent Written and Oral Communication Skills - My job as Student Assistant in the Office of Student Affairs has been in helping me to refine my communication skills, since it requires that I communicate daily with students and staff to clarify requests, provide information, and resolve problems in a professional manner.

3. Teamwork Experience - _____ University emphasizes on team projects, which has given me many opportunities to develop teamwork skills.

I will call you next week for the possibility of a personal meeting to discuss my qualifications. Thank you for your time and consideration.

第二封是向尚未公佈職缺的公司探聽目下或日後受雇的可能性：

Dear _____:

I have been working in high technology industry for the past five years and have followed your company with interest since its inception in June of 2000 It must be critical for a start-up company like yours to hire the most effective and appropriate person for each position and...

that is the reason why I am contacting you.

My experience includes a strong computer systems background with configuration, troubleshooting and support of Intel-based PCs and servers running Microsoft Windows series. As a result, you can trust my ability to bring knowledge of Intel-based computer programs as well as other technical skills such as video, communications and networking.

While there is not yet an open position posted in your Information Systems department, I do wish to express my sincere interest in your company. Any opportunity to meet with you to discuss current or future opportunities with your company will be welcome. I will call your office to arrange an appointment at your convenience.

Best regards,


　　有時候，向業界的的前輩求教，也可能發展出一個未來的工作機會，這個時候就需要一封有效的信來作敲門磚。下面這封請求會面和指教的信，大體上還可以，就是稍嫌囉唆了點。太長的信耗費讀者太多時間，對於比較忙的人，我們應該用比較精簡的方式來達意，尤其在有所求的時候。因此，原信可以改寫成右欄那樣精簡。

| 原信 | 修訂後 |
|---|---|
| Dear Mr. _____ :<br><br>I am currently a student in the MS Marketing Program of _____ University. My studies have exposed me to the multitude of career paths that can be traveled in the field of marketing. In the course of my exploration, I have become very interested in the area of consumer product marketing as a possible fit for my talents. My instructor, Mr. ____ __, suggested that I contact you to inquire about the possibility of arranging an informational interview to learn more about this exciting field from the point of view of an experienced and successful professional.<br><br>Evergreen Products is a respected leader in the consumer cosmetics industry, with a diverse clientele and a record of innovation in its product offerings and its approach to delivery of service. I would very much appreciate the opportunity to hear your thoughts about your own career, Evergreen Products, and the challenges and future opportunities that lie ahead in the area of consumer product marketing.<br><br>I would be very pleased to meet with you, at your convenience, for 20 minutes or whatever time you have available in your busy schedule. I will contact you in the next few days to see about arranging a meeting. Thank you for your consideration.<br><br>Yours truly, | Dear Mr._____ :<br><br>The opportunity to hear the thoughts of career, and the challenges as well as future opportunities in consumer product marketing from an experienced professional motivates me to write this letter.<br><br>As a student in the Marketing Department of _____ University, I am very interested in consumer product marketing. Dr. _____, my advisor, suggested me to contact you for the possibility of an informational interview to learn more about this field.<br><br>Meeting with you for 20 minutes or whatever time you have will be a great honor. I will contact you in the next few days for a possible arrangement. Thank you for your consideration.<br><br>Yours truly, |

　　向未曾謀面的人士要求會面、求教或請益，有時候是冒昧的事，但是，不入虎穴焉得虎子，最差的狀況不過是被拒絕而已。何況，對方也許恰巧被你那「初生之犢不畏虎」的精神所感動，讓你有意想不到的收穫。

## 二、推薦函（Recommendation）

　　選擇推薦人時，應該考慮對自己有基本認識、有好印象且願給你好評的人，師長、工作夥伴或以往的上司都是常見的人選。下面這封信可以用來請求推薦人寫推薦信：

Would you provide a letter of recommendation to support my job application and send it directly to Mr. _____ at _____? I recently applied for a _____ position there.

Please briefly describe any qualities which you feel would make me a productive employee in the _____ field, including any specific work you feel I did well on specific projects.

Thank you very much for your kindness.

　　如果可能，不妨請幾位推薦人各寫一封推薦信，複印備份，待需要時再行寄送，這樣可以避免一再麻煩這些人。

　　在人家需要幫忙的時候說些好話，是維持良好人際關係的很大助力，因此你若受到請託，不用吝惜這個舉手之勞，如果不知如何下筆，可以參考以下的推薦函例：

It is a great pleasure to recommend Ms._____ as a highly qualified _____. She worked for me for _____ years. During that time, she proved to be efficient, well organized and tactful with all those people she had to deal with.

I do not hesitate for a moment to recommend her.

下面這封是前上司提供的推薦信：

It has been a privilege to know Ms. _____ for _____ years in my role as _____ at _____ company.

While Ms. _____ reported to me, I found her management abilities to be invaluable in helping me to establish _____ Company as a leader in the market. Her conscientious effort and cooperation in doing professional, high-quality work were appreciated.

If you find that Ms. _____'s career objectives match your position description, I know of no reason you would be disappointed by her employment performance. Please let me know if you require further information.

底下這封則是以前的師長所提供的推薦函：

It is a pleasure to recommend my former student, Ms. _____, for employment with your firm.

As a student, Ms. _____ is efficient, innovative, and responsive. She motivates her peers with challenge and the opportunity for personal growth.

Ms. _____'s performance on the academic work has been excellent. As such, I am glad to recommend her to work with you.

推薦信不需耗費太多篇幅，重要的是誠懇具體，以上的幾封都達到了這種要求。

至於中文推薦函，下面的例子就很值得推薦：

各位先進：

很榮幸為貴校推薦_____先生。_____先生旅美近_____載，在彼邦的環保業界服務逾_____年，其間泰半在管理階層任職，所學所用皆與_____管理息息相關，具有豐富的實務經驗，實在是今日臺灣_____管理學界所亟需的教育人才。

_____先生為人和善，與同仁們極為相得，同時教學認真，廣受學生愛戴，他的閱歷之豐富，從他所能任教的科目之多可見一斑。_____先生富有創意，中英學養均為上駟，他雖所學的是理工，做的是管理，但他的人文素養較許多學文史哲者也不遑多讓，深信他在許多職務上都能勝任愉快，因此我毫不保留地向貴單位推薦_____先生。

如果各位對此信的描述有任何疑問，請隨時賜電_____。謝謝！

_____大學_____學系

_____謹識

## 三、謝函（Thank-you Letter）

　　面談之後即時向主持者去函道謝，除了是禮貌的表現之外，還可以博取很好的印象。謝函不須太長，只要能衷心表達謝意就夠了。

　　像左欄這封謝函就寫得長了些，而且大部分的篇幅是在提醒對方別忘了我的優點，這封信能得到的好感可能有限。此信的第二段比較適合出現在應徵函上，而不是出現在謝函裡，因此應該大幅刪減，修改後的右欄這封信，長度不到原信的四分之一，但致謝的味道更爲明顯。

| 原信 | 修訂後 |
|---|---|
| Dear Ms _____: | Dear Ms _____: |
| Thank you for allowing me the opportunity to interview for the _____ position. As the result of our informative discussion, my interest in the position has been strengthened substantially. | Thank you for a most enjoyable interview. |
| As I mentioned during our interview, I can bring thirteen years of experience in positions which required flexibility and the ability to accept and follow through on new assignments and responsibilities. I have had the opportunity to deal with people on all levels in the workplace as a Software Support Specialist and a Programmer Analyst. | Your informative briefing has strengthened my interest in the _____ position substantially. |
| Based on my past work experience, as well as my education, I feel confident that I have the qualifications for the position under discussion. Please let me know if there is anything I can do to assist you further in your hiring process. | Please call if you need additional information about my qualification. |
| Thank you for a most enjoyable interview. | Yours truly, |
| Yours truly, | |

下面這封謝函算得上得體，只是其中用的 *I* 還是稍多。若把原信改成右欄那樣，就更理想了。

| 原信 | 修訂後 |
|---|---|
| Dear Ms. _____: | Dear Ms. _____: |
| It was a sincere pleasure making your acquaintance Thursday regarding the position of _____. The knowledge I gained during the interview has certainly enhanced my interest in joining _____. | It was a sincere pleasure making your acquaintance during the interview regarding the position of _____. The information you provided has certainly enhanced my interest in joining _____. |
| I would also like to reiterate my interest in this position. I feel that it would be an exciting opportunity and believe my track records shows I would be a successful candidate. | This position is an exciting opportunity. You can be assured that my track record will make me a successful candidate. |
| I am looking forward to hearing your final decision. Thank you again. | Hope to hear about your favorable decision soon. Thank you again. |
| With best regards, | Yours truly, |

有個可以參考的原則是，信中出現 *I*、*My*、*We* 和 *Our* 的次數不要多過 *You* 和 *Yours* 的次數，而且不要在每段的開頭都用 *I*。

看看左欄的例子就可依照 You-ness 原則[1] 把信簡化成右欄那樣，把中心導向受信人。

---

[1]　見表 1-1「書信寫作原則」。

| 原信 | 修訂後 |
|---|---|
| Dear Mr. _____ : | Dear Mr. _____ : |
| Thank you for interviewing me yesterday for the _____ position. I enjoyed meeting you and learned more about _____ Publishing. | Thank you for the interview on ___ ____. Meeting you to learn the incredible opportunity offered by this _____ position is a great pleasure. |
| This position offers an incredible opportunity. I am eager to make a significant contribution to your corporation. | It will surely be wonderful working with you to make a significant contribution to _____ Corporation. |
| I have strong interest in the position and in working with you. Please feel free to call me at _____ if you need any additional information. | Please feel free to call me when you need additional information. |
| Sincerely, | Sincerely, |

再看一 "I" 個不停的例子和修改後的樣子：

| 原信 | 修訂後 |
|---|---|
| Dear Mr. _____ : | Dear Mr. _____ : |
| Thank you for taking the time to interview me today for the _____ position. I know that I am one of many who are being considered to work with your organization. I appreciated the opportunity to meet with you. | Thank you for taking the time to interview me today for the _____ position. |
| I have been looking for such a position and I am excited about the possibility of working for you. I am confident that I have both the experience and skills to be an asset to your company. | Knowing that my experience and skills can be an asset to your company and that there is a possibility to work for you is exciting. Hope to hear from you soon. |
| Thank you again for the interview. I look forward to hearing from you. | Sincerely, |

　　去函向爲你提供推薦信的人道謝是理所當然的事，請看這一封：

> Dear Mr. _____:
>
> Thank you very much for taking your time to write such a wonderful letter of recommendation. It will certainly have a very positive impact on any job application.
>
> Hope you don't mind my making copies to use it at similar occasions in the future. I will keep you informed about the outcome of my job application.
>
> Gratefully,

　　謝函可以用正式的郵件寄送，也可以借助 e-mail，受到幫助而知所道謝，日後如果再有所求，也會比較容易開口。

## 四、接受與婉謝函

　　雇主決定聘用後，你不論應聘與否，都要知會對方。知會的目的在讓對方的聘任作業保持順利。

　　同意應聘的信（Accepting Letter）在讓對方知道自己了解聘用的協定，也在表達謝意，左欄這封應聘函力求達到這兩點，但是重複之處稍多了些。簡短扼要外加親切有禮是寫作任何信件不變的指導原則，因此你可以刪去原信中的贅述，以 *You* 爲主角改進這封信（如右欄）。

| 原信 | 修訂後 |
|---|---|
| Dear Mr. _____ : | Dear Mr. _____ : |
| I am writing to confirm my acceptance of your employment offer of _____ and to tell you how delighted I am to be joining _____. The work is exactly what I have prepared to do. I feel confident that I can make a significant contribution to the corporation, and I am grateful for the opportunity you have given me. | I am glad to accept your offer of _____. This will also confirm my acceptance of the compensation and benefits package as described in your offer letter. You can expect me at the new employee orientation at 8:00 a.m. on _____. As you instructed, I will have completed the medical examination and all other compliance forms by that date. |
| I will report to work at 8:00 a.m. on _____. As we discussed, I will have completed the medical examination and all other compliance forms by that date. I will look forward to the new employee orientation. This will also confirm my acceptance of the compensation and benefits package as described in your letter. | It is exciting to have the opportunity to work with you and your fine team. Thank you for your confidence in me. |
| I am excited to have the opportunity to work with you and your fine team. I appreciate your confidence in me and am pleased to be joining your staff. | Yours truly, |
| Yours truly, | |

　　一旦不願或無法應聘，千萬不可形同陌路，一定要及早告知對方，讓對方及時另請高明，以免耽誤業務。這種告知的任務可以交給類似下列的婉謝函（Declining Letter）來完成：

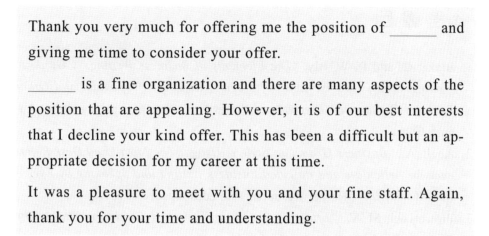

Thank you very much for offering me the position of _____ and giving me time to consider your offer.

_____ is a fine organization and there are many aspects of the position that are appealing. However, it is of our best interests that I decline your kind offer. This has been a difficult but an appropriate decision for my career at this time.

It was a pleasure to meet with you and your fine staff. Again, thank you for your time and understanding.

　　在婉謝接聘的信中,只要禮貌地說明一個簡單的理由,那個理由甚至不用太具體,這種信的目的主要在於知會,道歉並非絕對必要。

# 參考書目

1. Blake, G. and R. W. Bly, "*The Elements of Business Writing: A Guide to Writing Clear, Concise Letters, Memos, Reports, Proposals, and Other Business Documents (Elements of Series)*", Longman; reprint edition, August 1990

2. Abell, A., "*Business Grammar, Style & Usage: The Most Used Desk Reference for Articulate and Polished Business Writing and Speaking by Executives Worldwide*", Harper Resource, 2nd edition, July 1990

3. Piotrowski, M. V., "*Effective Business Writing: A Guide for Those Who Write on the Job*", Aspatore Books, April 2000

4. Cunningham, H. and B. Greene, "*The Business Style Handbook: An A-to-Z Guide for Writing on the Job with Tips from Communications Experts at the Fortune 500*", McGraw-Hill, 1st edition, February 2000

5. Davidson W., "*Business Writing: What Works, What Won't*", St. Martin's Griffin; revised edition, May 2000

6. Roddick H., "*Business Writing Makeovers: Shortcut Solutions to Improve Your Letters, E-Mails, and Faxes*", Adams Media Corp., November 2000

7. Danziger, E., "*Get to the Point! Painless Advice for Writing Memos, Letters and E-mails Your Colleagues and Clients Will Understand*", Three Rivers Press, December 2000

8. 葉乃嘉，《商用英文的溝通藝術》，新文京出版公司，2002/12。

9. 葉乃嘉，《知識管理》，全華科技圖書，2004/3。

10. 葉乃嘉，《中英論文寫作綱要與體例》，五南圖書出版公司，2005/1。

11. 葉乃嘉，《知識管理實務、專題與案例》，新文京出版公司，2005/8。

# 附 錄

# 長字化簡表

| 長字 | 簡潔代用字 | 中譯 |
| --- | --- | --- |
| above-mentioned | this | 上述的 |
| abundance | plenty | 豐富；充足 |
| acquire | get, gain | 獲得；取得 |
| apprise | inform, tell | 通知；告知 |
| advantageous | helpful | 有利的 |
| advise | say, tell | 勸告；告知 |
| aforementioned | this, these | 上述的 |
| alternative | another course | 二擇一 |
| anticipate | expect, look for | 預期；期望 |
| approximate | about | 近似；接近 |
| ascertain | find out | 查明；確定 |
| assistance | help, aid | 援助；幫助 |
| attitude | feeling | 意見；看法 |
| available | ready | 可利用的 |
| commence | begin, start | 開始；著手 |
| competent | able | 有能力的 |
| compliance with | according to | 合於 …… |
| comprised | made up of | 包括；包含 |
| concerning | about | 關於 |
| conclusion | end | 結論；結束 |
| confined | limited | 被限制的 |

| 長字 | 簡潔代用字 | 中譯 |
|---|---|---|
| cursory | short | 疏忽的；草率的 |
| demonstrate | show | 顯示 |
| deprive | keep from | 剝奪；阻止 |
| desire | want | 慾望；渴望 |
| determine | decide, find out | 裁定 |
| difficult | hard | 困難的 |
| effectuate | cause | 原因；導致 |
| efficacy | effectiveness | 有效 |
| efficient | able | 有能力的 |
| eliminate | leave out, omit | 消除；除去 |
| elucidate | explain | 解釋 |
| employ | use | 應用 |
| endeavor | try | 努力 |
| esteemed | respected | 受尊敬的 |
| eventuate | happen | 發生 |
| exceedingly | very | 非常地 |
| expedite | hasten | 促進 |
| explicit | clear, plain | 明確的；清楚的 |
| exterior | outside | 外部 |
| fabricate | make | 做；製造 |
| facilitate | help | 幫助；有助於 |
| furnish | give | 供應；給與 |
| henceforth | after this | 今後 |
| impact | affect | 影響 |
| inform | say, tell | 告知 |

| 長字 | 簡潔代用字 | 中譯 |
|---|---|---|
| initial | first | 最初的 |
| initiate | begin, start | 開始 |
| institute | begin | 開始 |
| interior | inside | 內部 |
| lenient | easy | 寬厚的 |
| necessitate | need | 需要 |
| negligent | careless | 粗心的 |
| participate | share | 參與 |
| perform | do | 做 |
| peruse, perusal | study | 閱讀 |
| possess | have | 擁有 |
| preserve | keep | 保持 |
| previous | before, former | 先前 |
| procure | get | 獲得 |
| prominent | leading | 突出的 |
| provide | give | 提供 |
| pursuant | according to | 根據；依照 |
| quantify | measure | 量化；測量 |
| reconcile | make agree | 使符合 |
| relinquish | give up | 放棄 |
| render | give | 給與 |
| request | ask | 要求 |
| require | need | 需要 |
| restrict | limit | 限制；約束 |
| retain | keep | 保持 |

| 長字 | 簡潔代用字 | 中譯 |
|---|---|---|
| secure | get | 獲得 |
| solution | answer | 解答 |
| sufficient | enough | 足夠的；充足的 |
| terminate | end | 結束；終止 |
| transpire | happen | 發生 |
| ultimate | final, last | 最終的；最後的 |
| utilize | use | 使用 |
| utilization | use | 使用 |

# 舊詞口語化表

| 舊式書信用詞 | 解釋 | 口語化後的用詞 |
|---|---|---|
| after which time | 然後 | then |
| As per the above label | 上面的地址 | To the address label given above |
| as per your letter, of the 15th at hand | 你（某日）的信上 | in your letter of June 15 |
| at a later date | 日後 | later |
| at an early date | 及早 | soon |
| at your earliest convenience | 儘快 | as soon as you can |
| Attached hereto is | 附上 | enclosed are 或 here are |
| by return mail | 立即 | promptly 或 immediately |
| Contents duly noted | 收悉 | We have read |
| Do not hesitate to | 請勿遲疑 | Please |
| Due to the fact that | 由於 | Because |
| Enclosed herewith and attached here to | 附上 | Enclosed are 或 here are |
| Enclosed please find | 附上 | Enclosed are 或 here are |
| First of all 或 first and foremost | 首先 | First |
| For your perusal | 請你詳閱 | Please read |
| Hereafter and henceforth | 此後 | From now on 或 In the future |
| I take the liberty to inform you | 很高興告訴您 | You will be glad to know |
| I wish to say that 或 I wish to state that | 謹告 | 刪除 |

| 舊式書信用詞 | 解釋 | 口語化後的用詞 |
|---|---|---|
| In accordance with your kind wishes | 據您要求 | As you requested |
| In accordance with your request | 據您要求 | As you requested |
| In due course of time | 即時 | In time |
| in the amount of | 數量；面值 | for |
| in the event that | 假如 | if |
| in view of the fact that | 因為 | since 或 because |
| inasmuch as 或 due to the fact that | 因為 | since 或 because |
| inform | 告知 | tell 或 say |
| Kindly advise the undersigned | 告知 | Please let me know |
| Kindly advise which design it is that you desire | 你想要什麼設計 | What design do you want |
| Kindly command me | 告知 | Please tell me |
| parsimonious | 吝嗇 | stingy |
| patronage | 惠顧 | business / order |
| Permit me to say | 容稟 | 刪除 |
| Pursuant to your recent request | 據您要求 | as you requested |
| take pleasure in | 很高興 | we are glad |
| Thank you kindly | 謝謝你 | thank you |
| the above-mentioned / the subject | 上述 | this |
| the enclosed self-addressed envelope | 附上回郵信封 | the enclosed envelope |
| the undersigned/to the writer/ yours truly | 我 | I / me |
| This is to inform you that | 謹告 | 刪除 |

| 舊式書信用詞 | 解釋 | 口語化後的用詞 |
|---|---|---|
| to my attention | 給我 | to me |
| under separate cover | 分別寄上 | separately |
| under the aforementioned circumstances | 在此情形下 | under these circumstances |
| We are contemplating | 我們正在考慮 | We are thinking |
| We are herewith changing your address | 你的地址已更正 | We've changed your address |
| We are pursuing a policy of | 根據 | According to |
| We are today in receipt of | 我們收到 | We receive |
| We note your request for | 你要求 | You ask for |
| We trust this will meet with your approval | 希望你能同意 | We hope you will approve |
| We want（或 would like）to thank you for | 謝謝你 | Thank you for |
| Will（或 Would）you be good enough to | 敬請 | Please |
| With your kind permission | 請准予 | Please |
| Would you kindly be good enough to send me | 請寄下 | Would you please send me |
| Your esteemed communication | 你的來信 | Your letter |
| Your letter of recent date | 你（某日）的來信 | Your letter on DATE |

# 贅詞簡化表

| 累贅的用法 | 簡潔的用法 | 解釋 |
|---|---|---|
| a decreased amount of | less | 減少的 |
| a decreased number of | fewer | 較少的 |
| absolutely essential | essential | 必要的 |
| accounted for by the fact | because | 因為 |
| adjacent to | near | 接近 |
| along the lines of | like | 類似 |
| an adequate amount of | enough | 足夠的 |
| an example of this is that | for example | 例如 |
| an order of magnitude faster | ten times faster | 10 倍速 |
| are of the same opinion | agree | 意見一致 |
| as a consequence of | because | 因為 |
| as a matter of fact | in fact（或略去） | 事實上 |
| as a result of | because | 因為 |
| as is the case | as happens | 同樣發生 |
| as of this data | today | 今天 |
| as to | about（或略去） | 關於 |
| at a rapid rate | rapidly | 立即 |
| at an earlier date | previously | 以前 |
| at an early date | soon | 不久；很快 |
| at no time | never | 從未 |
| at some future time | later | 以後 |
| at the conclusion of | after | 在 ……之後 |
| at the present time | now | 現在 |

| 累贅的用法 | 簡潔的用法 | 解釋 |
|---|---|---|
| at this point in time | now | 現在 |
| based on the fact that | because | 因為 |
| because of the fact that | because | 因為 |
| by means of | by, with | 以 ……方法 |
| causal factor | cause | 原因；導致 |
| cognizant of | aware of | 察覺 |
| completely full | full | 充滿 |
| consensus of opinion | consensus | 一致 |
| contingent upon | dependent on | 依 ……而定 |
| definitely proved | proved | 證明 |
| despite the fact that | although | 雖然 |
| determination is performed | is determined | 決定 |
| due to the fact that | because | 因為 |
| during the course of | during, while | 在 ……期間 |
| during the time that | while | 在 ……期間 |
| enclosed herewith | enclosed | 隨 ……附上 |
| end result | result | 產生；結果 |
| endeavor | try | 試 |
| entirely eliminate | eliminate | 排除；消除 |
| fatal outcome | death | 死亡；致命 |
| fewer in number | fewer | 較少的 |
| first of all | first | 第一 |
| for the purpose of | for | 為了 |
| for the reason that | because | 因為 |
| forthwith now | at once, | 立即 |

| 累贅的用法 | 簡潔的用法 | 解釋 |
|---|---|---|
| from the point of view of | for | 依……之見 |
| future plans | plans | 計畫；方案 |
| give an account of | describe | 描寫；描繪 |
| give rise to | cause | 原因；導致 |
| has engaged in a study of | has studied | 研究 |
| has the capability of | can | 能 |
| have the appearance of | look like | 看來像…… |
| having regard to | about | 關於 |
| his the investigators who | he | 他們 |
| implement | start | 開始 |
| important essentials | essentials | 必要的 |
| in a number of cases | some | 有些 |
| in a position to | can, may | 能；也許 |
| in a satisfactory manner | satisfactorily | 令人滿意 |
| in a situation in which | when | 在……時 |
| in a very real sense | in a sense（或略） | 就……來說 |
| in case | if | 假如 |
| in close proximity to | close, near | 關閉；近的 |
| in connection with | about, | 關於 |
| in light of the fact that | because | 因為 |
| in many cases | often | 時常 |
| in order to | to | 為了 |
| in relation to | toward, to | 關於；朝向 |
| in respect to | about | 關於 |
| in some cases | sometimes | 有時 |

| 累贅的用法 | 簡潔的用法 | 解釋 |
|---|---|---|
| in terms of | about | 關於 |
| in the absence of | without | 沒有 |
| in the event that | if | 假如 |
| in the not-too-distant future | soon | 不久；很快 |
| in the possession of | has, have | 擁有 |
| in this day and age | today | 今天 |
| in view of the fact that | because | 因為 |
| inasmuch as | for, as | 鑑於 |
| incline to the view | think | 想；思索 |
| is defined as | is | 是 |
| is desirous of | wants | 想要 |
| it is apparent that | apparently | 顯然 |
| It is clear that | Clearly | 清楚 |
| it is crucial that | must | 必須 |
| it is doubtful that | possibly | 可能的 |
| it is evident that A produced B | A produced B | A 產生 B |
| it is generally believed | many think | 一般認為 |
| it is of interest to note that | 略去 | 重點在 |
| it is often the case that | often | 時常 |
| It is worth pointing that | note that | 注意到 …… |
| it may, however, be noted that | but | 但是 |
| it should be noted that | note that（或略） | 注意到 …… |
| It was reported by _____ | _____ reported | Dr._____ 報告 |
| join together | join | 結合 |
| lacked the ability to | couldn't | 不能 |

| 累贅的用法 | 簡潔的用法 | 解釋 |
|---|---|---|
| large in size | large | 大的 |
| let's make it perfectly clear | 略去 | 理清楚 |
| make reference to | refer to | 提到；談論 |
| met with | met | 符合 |
| militate against | prohibit | 禁止 |
| more often than not | sometimes | 有時 |
| needless to say | 連同後文略去 | 不用說 |
| new initiatives | initiatives | 倡議 |
| no later than | by | 在……之前 |
| of great practical importance | useful | 有用的 |
| of long standing | old | 長久以來 |
| of the opinion that | think that | 思索那個…… |
| on a daily basis | daily | 每日的 |
| on account of | because | 因為 |
| on behalf of | for | 代表 |
| on no occasion | never | 從未 |
| on the basis of | by | 基於…… |
| on the grouds that | because | 因為 |
| on the occasions in which | when | 當……時 |
| on the part of | by, for | 就……而言 |
| owing to the fact that | because | 因為 |
| place a major emphasis on | stress | 強調 |
| pooled together | pooled | 集中 |
| presents a picture similar to | resembles | 類似；像 |
| previous to | before | 在……之前 |

| 累贅的用法 | 簡潔的用法 | 解釋 |
|---|---|---|
| prior to | before | 在 …… 之前 |
| quite unique | unique | 唯一的 |
| rather interesting | interesting | 有趣的 |
| red in color | red | 紅色 |
| referred to as | called | 稱作 |
| regardless of the fact that | even though | 即使；雖然 |
| relative to | about | 關於 |
| resultant effect | result | 發生；產生 |
| reverse side | other side | 反面 |
| root cause | cause | 原因 |
| serious crisis | crisis | 危機 |
| should it prove the case that | if | 如果 |
| smaller in size | smaller | 較小的 |
| so as to | to | 以便 |
| subject matter | subject | 主題 |
| subsequent to | after | 在 …… 之後 |
| take into consideration | consider | 考慮 |
| the question as to whether | whether | 是否 |
| the reason is because | because | 因為 |
| the result seem to indicate | the result shows | 結果表示 |
| through the use of | by, with | 以 …… 方法 |
| to the fullest possible extent | fully | 完全 |
| unanimity of opinion | agreement | 一致同意 |
| until such time | until | 到 …… 為止 |
| very unique | unique | 獨特的 |

| 累贅的用法 | 簡潔的用法 | 解釋 |
|---|---|---|
| was of the opinion that | believed | 相信；信任 |
| ways and means | way 或 means | 辦法 |
| what is the explanation of | why | 為何 |
| with a possible exception of | except | 除……之外 |
| with a view to | to | 為了 |
| with reference to | about（或略去） | 關於 |
| with regard to | about（或略去） | 關於 |
| with respect to | about | 關於 |
| with the result that | so that | 結果 |
| within the possibility | possible | 可能的 |

# 本書重要中英詞彙對照表

| Acronyms | 縮寫 | Acronyms | 縮寫 |
|---|---|---|---|
| Action Verb | 行動詞 | Complimentary Closing | 信末敬辭 |
| Active voice | 主動語態 | Congratulatory Letter | 祝賀信 |
| Address | 地址 | Context | 文意 |
| Adjective | 形容詞 | Core Competencies | 核心能力 |
| Administrative Skills | 行政能力 | Counseling Skills | 輔導能力 |
| Adverb | 副詞 | Creative Skills | 創作能力 |
| Advertisement | 廣告 | Credit | 信用 |
| Appearance | 外觀 | Customer, Client | 顧客 |
| Autobiography | 自傳 | Date | 日期 |
| Body of Letter | 正文 | Declining Letter | 婉謝函 |
| Case | 案例 | Education | 學歷 |
| Central Theme | 中心主旨 | Emotional Intelligence, EQ | 情緒智商 |
| Client | 客戶 | Empathy | 同理心 |
| Closing | 結語 | Executive Summary | 主管摘要 |
| Collection Letter | 催討信 | Extreme Block Style | 齊頭式 |
| Collection Letter | 催帳信 | Financial Skills | 財務能力 |
| Collection Letter | 催款信 | Follow-up | 後續行動 |
| Communication | 溝通 | Follow-up Letter | 隨訪信 |
| Communication Psychology | 溝通心理 | Form Letter | 制式信 |
| Communication Skills | 溝通能力 | Format | 格式 |
| Complain, Complaint | 抱怨 | Format of Letter | 書信格式 |

| Acronyms | 縮寫 | Acronyms | 縮寫 |
|---|---|---|---|
| Friendly | 友善 | Negative Tone | 負面 |
| General Rule | 通則 | Notice Letter | 通告函 |
| Greeting | 問候信 | Notification letter | 通知信 |
| Greeting Cards | 賀卡 | Official letters | 公務書信 |
| Greetings | 問候語 | Opening | 開場 |
| Inquiry | 洽詢 | Order | 下單 |
| Inside AddressDelivery address | 收信地址 | Order Card | 訂購卡 |
| Job Application | 求職 | Orders | 訂單 |
| Job Application | 應徵函 | Paragraphing | 分段 |
| Layout | 版面 | Passive Voice | 被動語態 |
| Letter of Acceptance | 應聘函 | Personal Data | 個人基本資料 |
| Letter of Commendation | 讚揚信 | Personal Trait | 人格特質 |
| Letter of Condolence | 慰問信 | Position Desired | 應徵職務 |
| Letter of Congratulation | 賀函 | Positive | 正面 |
| Letter of Declination | 婉拒函 | Positive Tone | 肯定 |
| Letter of Recommenda-tion | 推薦函 | Preferred Salary | 期望待遇 |
| Letterhead | 正式信箋 | Reader | 讀者 |
| Letterhead | 信箋 | Referrals | 介紹人 |
| Logic | 邏輯 | Repeat | 複述 |
| Management Skills | 管理能力 | Reply | 回信 |
| Mood | 心情 | Reply Card | 回郵卡 |
| Negative | 否定 | Report | 報告 |

| Acronyms | 縮寫 | Acronyms | 縮寫 |
|---|---|---|---|
| Research Skills | 研究能力 | Superfluous Words | 贅詞 |
| Response | 覆信 | Teaching Skills | 教學能力 |
| Resume | 履歷 | Technical Skills | 技術能力 |
| Return of Goods | 退貨 | Template Letter | 樣板信 |
| Sales Letter | 銷售信 | Thank-you Letter | 謝函 |
| Salutation | 敬稱 | Theme | 主題 |
| Signature | 信末簽名 | Title | 職銜 |
| Signature | 簽名 | Work Experience | 經歷 |
| Specialization | 專長 | Writing Method | 寫作方法 |
| Spoken Language | 口語化 | Writing Principle | 寫作原則 |
| Superfluous Word | 贅字 | | |

# 索 引

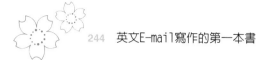

Note

Note

*Note*

國家圖書館出版品預行編目資料

英文E-mail寫作溝通的第一本書／葉乃嘉
作. 一 二版. 一 臺北市：五南, 2015.03
　　面；　　公分
ISBN 978-957-11-8024-3（平裝）

1.商業書信　2.商業英文　3.商業應用文
4.電子郵件

493.6　　　　　　　　104001749

1XN7　　研究方法與論文寫作

# 英文E-mail寫作溝通的第<br>一本書

作　　者 ─ 葉乃嘉(323.2)

發 行 人 ─ 楊榮川

總 編 輯 ─ 王翠華

企劃主編 ─ 黃惠娟

責任編輯 ─ 盧羿珊　周雪伶

封面設計 ─ 童安安

出 版 者 ─ 五南圖書出版股份有限公司

地　　址：106台北市大安區和平東路二段339號4樓

電　　話：(02) 2705-5066　　傳　真：(02) 2706-6100

網　　址：http://www.wunan.com.tw

電子郵件：wunan@wunan.com.tw

劃撥帳號：01068953

戶　　名：五南圖書出版股份有限公司

台中市駐區辦公室/台中市中區中山路6號

電　　話：(04) 2223-0891　　傳　真：(04) 2223-3549

高雄市駐區辦公室/高雄市新興區中山一路290號

電　　話：(07) 2358-702　　傳　真：(07) 2350-236

法律顧問　林勝安律師事務所　林勝安律師

出版日期　2005年9月初版一刷
　　　　　2015年3月二版一刷

定　　價　新臺幣350元